Indoor Environment and Health

Edited by Orhan Korhan

Published in London, United Kingdom

IntechOpen

Supporting open minds since 2005

Indoor Environment and Health
http://dx.doi.org/10.5772/intechopen.78905
Edited by Orhan Korhan

Contributors
Mahima Habil, Himanshu Dehra, Maher Shehadi, Takemi Otsuki, Young Lee, Igor Cretescu, Dorina Nicolina Isopescu, Doina Lutic, Gabriela Soreanu

First published in London, United Kingdom, 2019 by IntechOpen
IntechOpen is the global imprint of INTECHOPEN LIMITED, registered in England and Wales, registration number: 11086078, The Shard, 25th floor, 32 London Bridge Street
London, SE19SG – United Kingdom
Printed in Croatia

British Library Cataloguing-in-Publication Data
A catalogue record for this book is available from the British Library

Additional hard and PDF copies can be obtained from orders@intechopen.com

Indoor Environment and Health
Edited by Orhan Korhan
p. cm.
Print ISBN 978-1-78984-373-6
Online ISBN 978-1-78984-374-3
eBook (PDF) ISBN 978-1-78984-606-5

For EU product safety concerns:
IN TECH d.o.o., Prolaz Marije Krucifikse Kozulić 3, 51000 Rijeka, Croatia, info@intechopen.com or visit our website at intechopen.com.

Meet the editor

Orhan Korhan graduated with a BS from Eastern Mediterranean University in 2000, an MS from the University of Louisville in 2002, and a PhD from Eastern Mediterranean University in Industrial Engineering in 2010. He has been working at Eastern Mediterranean University since 2009. He became Assistant Professor in 2010 and Associate Professor in 2014. He is a board member of the Continuous Education Center and served as Academic Affairs Coordinator at the Rector's Office from 2016 to 2018. He has been assigned to the scientific committees of several international conferences and published several books, book chapters, and papers in various countries. His current research interests include work-related musculoskeletal disorders, educational ergonomics, human–computer interaction, and facilities planning and design.

Contents

Preface

It has always been a major concern for humans to find a safe and comfortable environment. Indoor environment quality became an important factor for health, comfort, and performance. The right balance of acoustic, thermal, and visual comfort together with good air quality improves health, wellbeing, workplace productivity, and learning outcomes. The concepts of heating, ventilation, air conditioning, external temperature, air pollution, sick building, indoor pollutants, illumination, glare, indoor lighting, daylight, noise, construction materials, sound intensity, and furniture on the indoor environment are described in detail. This book, therefore, helps to understand the impact of a sustainable healthy and comfortable indoor environment on the quality of life, and perceives the required indoor conditions for productivity and effectiveness.

Assoc. Prof. Dr. Orhan Korhan
Eastern Mediterranean University,
Cyprus

Section 1

Indoor Air Quality

Chapter 1

Air Quality and Airflow Characteristic Studies for Passenger Aircraft Cabins

Maher Shehadi

Abstract

This chapter summarizes the work done at the Airliner Cabin Environment Research Lab (ACERL) related to air quality, airflow characteristics, and human thermal comfort inside aircraft cabins. The laboratory is part of the Institute for Environmental Research (IER) at Kansas State University. It has a Boing 767 mockup cabin, bleed air simulator, and a Boeing 737 actual aircraft section that were all utilized to conduct experimental studies to understand air quality inside aircraft cabins. The studies summarized in this chapter include particle image velocimetry (PIV) investigations, particle dispersion, computational fluid dynamics (CFD) simulations, tracer gas and smoke visualization studies, and bleed air investigations. The chapter also summarizes other related studies including virus dispersion, air quality monitoring devices, and related developed air quality standards. The scope of this chapter is to summarize the setup and results of each of the above categories. This summary along with the cited references provides results for full size aircraft cabin environments, helps validate data for CFD simulations, and provides comparison data for other similar studies. This helps improve the design of future aircraft cabins and their ventilation systems and recommends changes to maintenance practices done that can improve the health and safety of humans inside these enclosed compartments.

Keywords: indoor air quality, airflow simulations, human health and safety, particle dispersion, aircraft cabin environment, bleed air, tracer gas

1. Introduction

Biological incidents such as severe acute respiratory syndrome (SARS) and swine flu (H1N1) transmission have been detected on flights. Ebola was the latest virus threat on board air-flights. Chemical incidents are also detected and reported inside commercial aircraft passengers' cabins. Odors and fumes from bleed air, viruses, and bacteria can result in serious health hazards for cabin crew members and passengers and have an important impact on aircraft air quality [1].

The purpose of this chapter is to summarize studies related to airflow studies, air speed and turbulence characteristics inside aircraft cabins. This can help aircraft manufacturers and operators in providing the occupants with acceptable air quality that meet safety guidelines and codes while maintaining the passengers' comfort in these enclosed and pressurized compartments.

In addition to that, the studies presented in this chapter provide important information to help in validating simulations and CFD codes developed to understand airflow behavior inside aircraft cabins.

2. Facilities and experimental setup

The Airliner Cabin Environment Research Lab (ACERL) houses two Boeing mockup cabins, a bleed air simulator, and a half cabin generic room. The details and specifications of each of the four structures are as follows:

2.1 Generic aircraft half-cabin mockup model

This generic model represents a half, twin-aisle, Boeing aircraft cabin. It has dimensions of L × W × H = 2.1 m × 2.1 m × 1.7 m in height (6.4 ft × 6.4 ft ×5.1 ft). The actual built-up room and a generic CAD model for this facility are shown in **Figure 1** [2, 3].

2.2 Boeing 767 cabin mockup

This structure represented a full size and wide body Boeing aircraft cabin mockup that was entirely built and validated at the ACERL. It consisted of 11 rows of seats, along the length of the cabin, with seven seats in the lateral (transverse) direction. The dimensions of the cabin mockup were 9.6 m (31 ft 5 in.) in length and 4.7 m (15 ft 6 in.) wide at the widest spot right above the arm seats area. The mockup cabin seats, the air supply duct, and linear diffusers are original parts from a salvaged Boeing 767 aircraft. Each seat in the cabin was occupied by an inflatable manikin, shown in **Figure 2c**, which was instrumented with a 10 m-long wire heater element to generate approximately 100 W (341 Btu/h) of distributing heat, representing heat gains from a sedentary human being. The supplied air to the cabin mockup was 100% fresh air with no recirculated air. Air temperature was controlled via a chiller and a heater loop as shown in **Figure 2b**. Prior to supplying the outdoor fresh air into the cabin mockup, a set of high-efficiency particulate air (HEPA) filters was installed in series with the supply duct, as shown in **Figure 2a**. Following the HEPA filter, the air was supplied to linear diffusers inside the cabin mockup. Air was exhausted from the bottom of the cabin mockup. The total airflow rate supplied to the cabin was approximately 660 l/s (1400 CFM) of fresh air, thus allowing 8.57 l/s (18 CFM) for every seat of the 77-seats available in the cabin. The

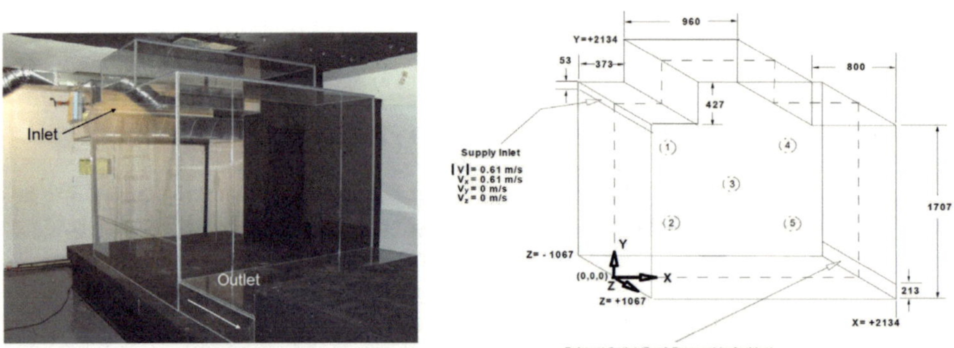

(a) (b)

Figure 1.
Half full scale cabin (a) generic room [2] and (b) generated CAD model [3] (dimensions in mm).

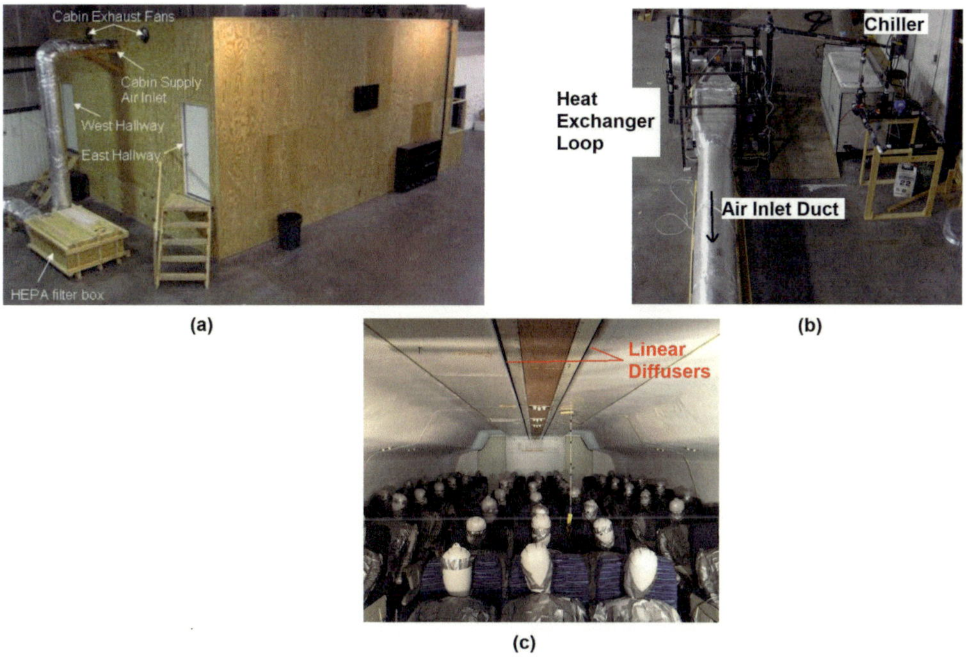

Figure 2.
Boeing 767 cabin mockup (a) exterior, (b) supply air system and control, and (c) interior of the cabin with inflatable manikins in each seat shown [1, 4, 5].

temperature of the air inside the cabin was controlled to 23–25°C which met the recommended design values for inside passenger aircraft cabins. It was important to insure that the air supplied through the two linear diffusers was balanced to maintain airflow uniformity inside the cabin. Also, the duct supplying the two diffusers had a decreasing diameter along the length of the cabin to allow uniform pressure. Although the uniformity of the air exiting the cabin was not a necessity, the exhaust area was uniform along the full length of the cabin and was open into a larger plenum that has negligible pressure gradients.

2.3 Boeing 737 sectional cabin

Two different sections of actual narrow body Boeing 737 aircraft cabins were used. The first one included three rows of seats with a total number of 18 seats. The seats were equipped with heated cylinders releasing approximately 100 W (341 Btu/h) except one seat that was equipped with a thermal manikin simulating a human body generating 100 W (341 Btu/h), as well. Cabin dimensions are shown in **Figure 3**. The other cabin size was 5.6 m (18.3 ft) in length, 3.6 m (11.8 ft) in width, 2.8 m (9.2 ft) in height and is shown in **Figure 4**. It consisted of five rows with 30 seats in total. All seats were equipped with similar inflatable-heated manikins as was described and used in the Boeing 767 cabin mockup. The airflow and temperature for both cabins were controlled meeting the specifications described for the Boeing 767 cabin mockup described previously.

2.4 Bleed air simulator

To help in identifying contamination sources onboard actual aircraft, HEPA filters were extracted from the air recirculation ducts from actual aircraft and were analyzed using the bleed air simulator shown in **Figure 5**. To simulate oil leakage,

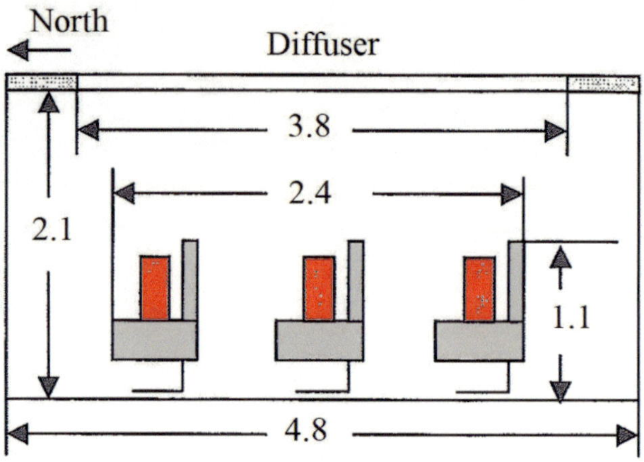

Figure 3.
3-rows B737 cabin side view schematic (dimensions in mm) [6].

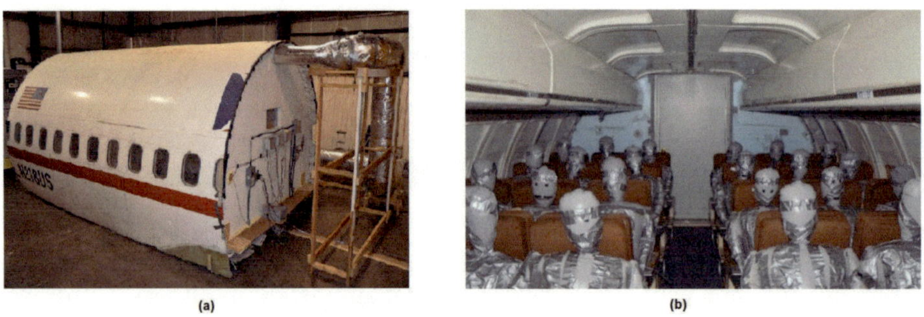

Figure 4.
5-rows B737 (a) exterior and (b) interior with inflatable heated manikins [7].

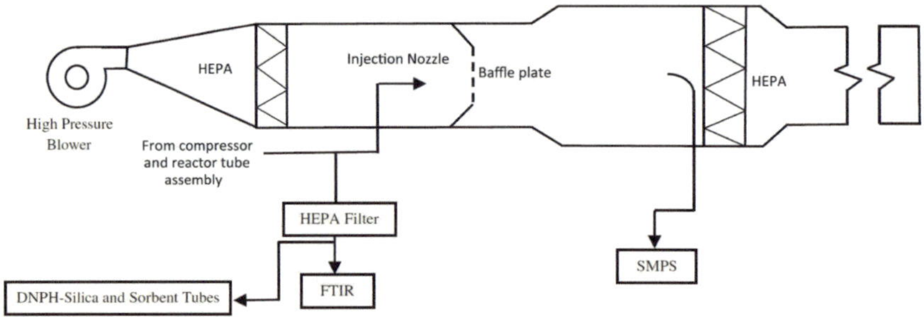

Figure 5.
Bleed air simulator [8].

a compressor was used with an oil generator forcing oil droplets in the testing duct where the filters were installed. A reciprocating compressor followed by a heated tube was used to create controlled temperature and pressure conditions representative of bleed air from an aircraft engine. Aerosolized lubricating oil was injected into the airflow upstream of the compressor and the particulate characteristics were measured downstream of the heated tube. Gas chromatograph-mass spectrometry (GC/MS) was used in the analysis [9].

Important key factors that can affect the accuracy of the results obtained were the duration of oil existence on the filters, storage conditions and duration of the filters, and oil evaporation [10].

3. Summary of studies

The progress of the studies conducted at the ACERL Lab is summarized as shown in **Table 1**. Investigation started with PIV techniques for structures' validation and to help in validating the computational fluid dynamics (CFD) codes development. This included multiple studies such as by [3, 6, 13, 16]. At the same time, literature was conducted to understand and evaluate air quality inside aircraft cabins such as by [11, 12, 14, 18]. In 2006, a new phase of investigation for air quality and airflow was approved by the US Federal Aviation Administration (FAA). The team started their investigation inside the B767 cabin mockup described in Section 2.2 and shown in **Figure 2**. The investigation started with particles dispersion inside the aircraft cabin mockup while working on more literature including air characterization inside aircraft cabins, contamination and viruses monitoring development, and other viruses related literature studies. After obtaining enough initial results,

	CFD	PIV	Tracer gas	Particles	Bleed air	Others
2001		Hosni and Jones [6]				
2002						Jones [11, 12]
2003	Hosni and Jones [13]					Jones [14]
2005	Lin et al. [3], Lebbin et al. [15]					
2006	Lin et al. [16], Lebbin et al. [17]	Lin et al. [16]				Jones [18]
2008				Padilla [19]		Jones [20, 21]
2010		Ebrahimi et al. [22]		Shehadi et al. [23]		Jones [24], Loo et al. [25]
2011			Trupka et al. [26, 27]	Beneke et al. [28]	Korves et al. [29]	
2013		Ebrahimi et al. [2, 30, 31], Isukapalli et al. [32]	Anderson et al. [33]			Korves et al. [34]
2014			Shehadi et al. [35]	Powell et al. [36]	Eckels et al. [9], Mann et al. [37]	
2015			Shehadi [4], Keshavarz et al. [38]			Shehadi et al. [39]
2016			Patel et al. [5]		Omana et al. [10]	Shehadi et al. [40]
2017			Patel et al. [7], Amiri et al. [8]		Jones et al. [41], Space et al. [42]	
2018					Shehadi et al. [1]	
2019				Shehadi [43]		

Table 1.
Studies done at the ACER Lab (indexed as in the reference section) and their corresponding study category.

different approaches were followed using CFD and tracer gas testing while continuing to work on particle dispersion investigations. At the same time, bleed air studies using the bleed air simulator shown in **Figure 5** were taking place to analyze the chemical composition of oil that penetrate through the aircraft circulation HEPA filters, to understand the behavior of oil under high temperatures and pressures and to help assess their impact on human health and comfort inside the aircraft. Beyond that, CFD, tracer gas, and bleed air investigation were all conducted in parallel while checking on available literature related to bacteria and virus outbreaks inside aircraft cabins and other fume and bleed air issues, as well. The reviewed studies are presented in a chronological order as presented in **Table 1**.

3.1 Particle image velocimetry (PIV) studies

The initial testing and verifications started using PIV techniques inside a generic half cabin model and sectional Boeing 737 as described in Sections 2.1 and 2.3 and shown in **Figures 1** and **3**. Hosni and Jones [6, 13] used the sectional narrow body Boeing 737 aircraft cabin, shown in **Figure 3**, to measure velocity and turbulence intensities inside a 3-row B737 airplane cabin. PIV techniques were used to meet these objectives. Each row consisted of six seats in the lateral (transverse) section of the cabin resulting in total 18 seats. A thermal manikin was used in one seat to simulate the human body inside the cabin and the remaining 17 seats were equipped with 100 W (341 Btu/h) cylindrical heaters simulating heat output of seated passengers. A total of nine planes were used for PIV measurements with individual measuring sections of 0.61 m × 0.61 m (2 ft × 2 ft) each. The results for airflow velocity and turbulence intensities were to be used to validate further developed CFD codes.

In 2005, two other studies were published. The first one included a comparison between computational results from large eddy simulation (LES) and experimental PIV results inside a generic cabin model. The cabin model dimensions were L × W × H = 2.1 m × 2.1 m × 1.7 m, as shown in **Figure 1**. LES predictions and PIV results were in agreement and so did the energy spectral analysis [3]. The second study by Lebbin et al. [15] compared results of airflow measurements using various measurement tools and equipment inside the same generic cabin model as shown in **Figure 1**. Some of the measurement tools include stereoscopic PIV (SPIV), sonic anemometer, hot-wire anemometer, and draft instrument. The study showed the benefit and significance of using a noncontact measurement system such as an SPIV system. Another study comparing LES with PIV for the airflow in the same generic cabin model was published in 2006 by Lin et al. [16]. The study came to the same conclusions as [3].

In 2006, Lebbin et al. investigated the effect of different nozzle sizes on the air inlet velocity inside the generic room described in **Figure 1** using SPIV technique. Reynolds number was held constant at the inlet slot of the room with a value of approximately 2226. It was noted that the center of rotation of the overall airflow significantly changed with a change in the size of the air inlet slot size, whereas the turbulence levels in the room was not affected significantly since Reynolds number was not changed [17].

3.2 Particle dispersion studies

Following PIV testing and before conducting any infectious and bacterial transport investigations, particulate transport and behavior inside aircraft cabins were studied. Padilla [19] investigated particulate transport in a half cabin generic room shown in **Figure 1**. Two different particle sizes were investigated: 3 and

10 μm. Each of the two particle sizes was investigated separately. Particles were injected at a constant point inside the room at a height of 606.6 mm (24 in.) above the floor on the centerline of the test room. Particles were measured at five different locations throughout the cabin. It was found that the normalized concentrations of 3 μm particles were close to one which confirmed the well air mixing inside the room. However, the 10 μm testing showed much lower normalized concentrations approximately 0.1. Several changes were done to the injection nozzle and cabin pressure and the results came to 0.4 and 1.5. Therefore, it was concluded that the 3 μm particles followed the airflow inside the cabin. A similar conclusion was difficult to reach for the 10 μm particles.

Particle dispersion inside an 11-row B767 cabin mockup was experimentally investigated by Shehadi et al. [23] to determine the best location for placing a particle detection sensor in the transverse direction of the B767 cabin mockup shown in **Figure 2**. The cabin mockup consisted of seven seats in the transverse direction with two linear diffusers around the center line of the cabin ceiling supplying air to the cabin mockup. Poly-disperse talcum powder, with a density of 0.95 g/cm^3, was used as the testing agent and was released to simulate a sneeze from a seated passenger. Pressurized air was instantly directed into the powder forming a powder cloud at a height equivalent to the nasal area of a seated passenger. It was concluded that a properly placed sensor can accurately detect released particles in the transverse direction of the cabin when released in the same row as well as one adjacent row ahead and behind the release-row. This was identified as near field detection. Another study investigating particles dispersion inside the B767 cabin mockup cabin was done by Beneke et al. [28]. The paper indicated that regions around the release source had the highest exposure of particles released and the highest variations, as well. This exposure and variation level followed an exponential decay along the length of the cabin from the release-source at the second row of the cabin mockup toward the back wall (far field detection). A recent study in the same Boeing 767 mockup cabin investigating particles between 05 and 5 μm showed that there is a 20–35% additional drop in normalized particle counts per row away from the release row in the longitudinal direction [43].

Powel et al. [36] investigated particle deposition rates inside the B767 cabin mockup shown in **Figure 2**. Particle deposition was investigated over multiple locations inside the cabin and was collected at different surface orientations, horizontally and vertically aligned. Particles were collected on clean tape surface and were optically counted using fine small grids under a photographic type microscope. It was found that the surface orientation played a significant role in particle deposition rates with approximately 1-order of magnitude difference between vertical and horizontal orientation.

3.3 Computational fluid dynamics (CFD) studies

Tracer gas, smoke visualization and particle testing were all done inside the generic room mimicking a half aircraft passenger cabin, inside actual cabin sections, and inside a cabin mockup to understand airflow and turbulence characteristics such as velocity and turbulence intensity. Each study was conducted with new and different goals. However, all the studies were intended to support the development and validation of mathematical and computational models for airflow and particles or gas dispersion in aircraft cabins.

Ebrahimi et al. [22] conducted several computational studies for the Boeing 767 cabin mockup, shown in **Figure 2**, investigating the airflow and turbulence characteristics and validating CFD models by comparing against available experimental data. Large Eddy Simulation (LES) and Reynolds Averaged Navier-Stokes

(RANS) methods were used for simulation in the study. Computational results were validated against previously presented data in the PIV studies section. Other CFD results from literature were used against the simulated results for further validation. Nozzle height effects from the cabin floor on the airflow behavior were investigated in the study. Results from LES with Werner-Wengle wall function were found to predict unsteady airflow velocity fields inside the Boeing 767 cabin mockup with relatively high accuracy. However, in places where air circulation took place and accuracy was not as good, the RGN k-ε model with the non-equilibrium wall function model predicted the steady-state airflow velocities with good agreements with experimental results [2]. In a later CFD study by Ebrahimi et al. [30], the calculated velocity was used to predict tracer gas and particle dispersion simulation inside the full-scale, 11-row, Boeing 767 aircraft cabin mockup shown in **Figure 2**. In the same study by Ebrahimi et al. [30], RANS method and the RGN k-ε model were used for airflow simulations and turbulence modeling, respectively. Three different grid sizes were examined for grid uncertainty analysis. To start the simulations, initial airflow velocities exiting the supply nozzles were experimentally measured using omni-probes. The boundary conditions were further refined by conducting continuous comparisons between experimental and simulated results.

CFD modeling inside the Boeing 767 cabin mockup was used to understand the pesticide deposition rates. Pesticides are usually used by many countries for disinfection purposes. Different spraying patterns with different angles were simulated at both low and high cabin air exchange rates. The pesticide deposition samples were collected at the lap and seat top levels. The developed models predicted results with high accuracy when simulating high air exchange rates but underestimated the concentrations at window seats when under low air exchange rates. No major variations were found in deposition characteristics between sideways and overhead spraying angles [32].

In 2011, while simulating different scenarios for the Boeing 767 cabin mockup, Ebrahimi et al. [31] conducted computational analysis for the generic half cabin room, shown in **Figure 1**, using the Lagrange-Euler approach. Air was modeled as continuous phase whereas particles were treated as discrete. The discrete and continuous phases were solved through the calculation of drag and buoyancy forces acting on particles. Reynolds Averaged Navier-Stokes (RANS) method was used for velocity calculations while checking the dependency of RANS simulations on grid size through a controllable local mesh refinement scheme. It was found that unstructured grid with tetrahedral and hybrid elements provided better results than structured grid with hexahedral elements [31].

3.4 Tracer gas studies

Following PIV, CFD simulations, and particle dispersion testing, tracer gas testing was used to investigate airflow speed, orientation, and turbulence characteristics. Tracer gas has been widely used in experimental studies to study ventilation effectiveness, airflow circulations, airflow velocities, and other parameters inside aircraft cabins, enclosed environments and structures, buildings, hospitals, and many other applications. Some of the properties of the tracer gases to reveal unbiased results are reactivity and sensibility, should be non-reactive, should not react chemically or physically with any part of the system under study, and should be insensible otherwise the results obtained from the processes under investigation would be biased. In other words, tracer gases shall not affect the airflow or air density of the system, should not change the state of air, its chemical properties, and should mix and follow the airflow well. In addition to that, the tracer gas used should have measurable criteria in order to be quantified. On top of all of the above properties, the used tracer gas shall be safe and should be non-flammable, non-toxic, and non-allergenic [4]. All of the tracer gas studies that will be presented here-in were conducted inside the Boeing 767 cabin mockup described in **Figure 2**, except the study done by Patel et al. [7].

Smoke visualization techniques along with tracer gas, mainly made up of carbon dioxide and mixed with helium to maintain equivalent buoyancy with air in the cabin, was used to understand airflow behavior inside the Boeing 767 cabin mockup described in **Figure 2**. The tracer gas was released and sampled randomly throughout the cabin mockup. Smoke visualization results showed that the flow inside the cabin was chaotic and difficult to quantify. Quantitative results obtained from tracer gas testing helped in identifying several swirling and circulations inside the cabin. Some circulations were clockwise and others were counter clockwise directed. The circulations inside the cabin mockup were believed to be part of the driving mechanism in transporting tracer gas through the longitudinal direction of the cabin [35]. More analysis was done by Shehadi et al. [1] to analyze airflow and turbulence characteristics inside the same Boeing 767 cabin mockup. Tracer gas sampling was used as the main technique. Several eddies and circulations inside the cabin were identified with two large-size circulations dominating over the front and middle sections of the cabin mockup. The airflow in the aft section of the cabin was shown to be more chaotic and experienced more complex flow characteristics than in other sections of the cabin. Uncertainty calculations were done to check on the measurements' accuracy and validation and it was approximately ±14% including bias and random uncertainties. Also, it was concluded that the longitudinal length of the cabin controls the number of air circulations that could be present in cabins of similar aircraft model and type.

In another study, carbon dioxide was used as a tracer gas to simulate a gaseous decontamination agent inside the Boeing 767 mockup cabin shown in **Figure 2**. Supplementary axial fans were used in both and either aisles of the cabin mockup in separate cases and scenarios. Geometric, thermal, and airflow boundary conditions were also measured and documented. Gaseous transport inside the cabin was shown to be non-symmetrical between the two transverse sections of the cabin at several locations although the cabin and boundary conditions were symmetrical. This asymmetry was due to horizontal circulations of air that naturally formed in the cabin. This circulation had a substantial impact on longitudinal gaseous transport [38].

Tracer gas made up of carbon dioxide was used to investigate the effect of a moving cart by a cabin crew in one of the aisles inside the Boeing 767 aircraft cabin mockup described in **Figure 2**. Tracer gas was injected inside the cabin at constant rates and then was quantitatively sampled using non-dispersive infrared (NDIR) sensors at different locations inside the cabin. The impact of the moving cart and the attached dummy, which represented a cabin crew, was insignificant compared to the transport via cabin air motion [26, 27]. Most of other variables and disturbances in airliner cabins related to the impact of a beverage cart and the cabin crew driving it do not appear to provide a significant path for longitudinal dispersion of contaminates.

The effect of gaspers or personal air outlets used in aircraft cabins on the airflow and transport phenomena inside the Boeing 767 cabin mockup was investigated by Anderson et al. [33]. Tracer gas was released and sampled in the breathing zone of a seated passenger using a thermal manikin. While keeping the airflow supply at constant rate, personal supply gaspers were found to impact local exposure by disrupting the contaminant plume. In some cases, there was significant reduction in close-range, person-to-person exposure, while in other cases there was negligible or even negative impacts. No concrete conclusions or universal guidelines could be identified for the use of gaspers due to the unpredictable behavior of the plumes and due to variations in their behavior from location to location. However, it was found in most cases that it was more effective to use the gaspers by the source person rather than by the exposed person as the air leaving the gasper nozzles tended to push the contaminant plumes from the source person down and out of the breathing zone [33].

Local ventilation effectiveness inside the B767 cabin mockup and the 5-rows Boeing 737 cabin section described in **Figures 2** and 4, respectively, was investigated by Patel et al. in [5, 6], respectively. Experiments inside both cabin models were conducted using tracer gas. Tracer gas, composed mainly of carbon dioxide, was sampled at all of the 77 seats inside the 11-row Boeing 767 cabin mockup. The overall ventilation rate was found approximately at 27 air changes per hour (ACH) based on total supply air flow. The ventilation effectiveness ranged from 0.86 to 1.02 with a mean value of 0.94. These ventilation effectiveness values were higher than what typically is found in other indoor environments. This gain in effectiveness is likely due to the relatively high airspeeds that can improve mixing rates [5]. On the other hand, experiments inside the 5-rows sectional Boeing 737 were carried with similar thermal manikins as was used in the Boeing 767 cabin mockup [7]. Local ventilation effectiveness was found to be uniform throughout the Boeing 737 cabin regardless of the location inside the cabin. For gaseous transport, similar conclusions were found in both cabins, Boeing 767 and Boeing 737, where gaseous transport was significantly transported in the transverse and longitudinal directional of the cabins.

3.5 Bleed air characterization studies

Bleed air studies were done to investigate bleed air contaminants generated during incidents such as fume or oil thermal degradation incidents inside aircraft cabins. Multiple devices and different techniques were used to investigate bleed air contamination inside aircraft.

Contamination source identification is very difficult and would require multiple occurrences inside aircraft cabins [10]. Eckels et al. [9] recommended analyzing the high-efficiency particulate air (HEPA) filters used with the air recirculation ducts of aircraft in nearly all commercial aircraft. This procedure and technique would provide a chemical map database or library for contamination particles found on HEPA filters and ultimately inside aircraft cabins. A bleed air simulator was built for this purpose and is shown in **Figure 5**.

Gas chromatography/mass spectrometry (GC/MS) was found useful in providing information about the likely contamination source, but further research was deemed necessary to validate the methods. For example, the effect of oil existence duration on the filters can change the whole conclusions. If oil contamination on HEPA filters is not time specific, then if it has occurred recently or weeks or months prior to the testing of the filters, it would not have much difference. It could be the result of continuous low-level contamination or it could be the result of one or a few significant events. Knowledge of the stability of the oil on HEPA filter media is thus crucial for the validation of GC/MS testing and analysis [10]. Air filter sampling and the ResPlex II assay were found to be an effective method in identifying and characterizing viruses in aircraft cabins' air by analyzing the cabin filters [29].

Results from [9] in the bleed air simulator, shown in **Figure 5**, when investigating standard and nonstandard filters, showed a concrete link between tricresyl phosphates (TCPs) and a homologous series of synthetic pentaerythritol esters from oil and contaminants. High correlation was found for nonstandard filters than with standard ones.

Mann et al. [37] compared results of four particle counters used inside the bleed air simulator that was described in Section 2.4 and shown in **Figure 5**. The particle counters included a scanning mobility analyzer, an aerodynamic particle-sizer, an optical particle counter, and a water-based condensation particle counter. The covered particle sizes ranged between 13 nm and 20 μm. Effects of temperature and pressure on particles generation were investigated. It was found that high temperatures can increase ultrafine particles existence, whereas the pressure of the bleed air had little discernible effect on both particle size and concentration.

Another study analyzing the concentration, number of counts and size of different chemicals was done by Amiri et al. [8]. Various temperatures and pressures were considered inside the bleed air simulator described previously. The results showed that different aldehydes were formed with increasing concentrations with pressure and temperature. Carbon-monoxide was noticed to increase in concentration with both increasing pressure and temperature across all temperatures and pressures evaluated. It was noticed that the minimum bleed-air temperature resulted in maximum particle sizes and minimum concentrations.

To check on the compliance air quality with aircraft codes and design guidelines, specifically the ANSI/ASHRAE Standard 161, in detecting contamination from lubricating oil, a four-part experimental program was done by Jones et al. [41]. For the first part, the bleed air simulator was used to check on the temperature and pressure effects. For the second and third parts, turbine shaft engines mounted in test stands were used. The fourth part of the program was done on a NASA Vehicle Integrated Propulsion Research (VIPR) study which was conducted on an US Air Force C-17 Globemaster III military transport aircraft. Particulate size distributions and concentrations were measured with aerodynamic particle sizer and scanning mobility particle sizer, as well. Very low contamination rates were found. However, it was noted that many of the droplets may be even smaller than 10 nm which raises the necessity for developing ultrafine particle detection and sensing of low contamination levels. The results for the VIPR program showed that chemical contaminants from the injected engine oil could be captured by various types of sampling media. After analysis, it was found that no significant concentrations of contaminants in bleed air exceeded established OSHA PEL and STEL limits [42].

3.6 Other studies

Other studies included developing air quality standards such as by [14, 18, 20, 21] and summarizing data from literature and previous studies that can help in regulating air quality inside passenger aircraft [11]. Other studies investigated advanced models for predicting the transport of infectious disease, viruses and contamination in airplanes [24] and analyzed bacteria development on airplane HEPA filters that can facilitate the design of biosensors for infectious organisms' detection on commercial aircraft [34].

A study by Loo and Jones [25] discussed the requirements for developing a portable unit that can detect and measure air quality on board actual flights. Some of the recommendations for a good and reliable sensor unit were: (1) cost should be less than $1000 for each unit after development is completed, (2) certified for electromagnetic interference (EMI) that would allow it to operate during all phases of a given flight, (3) no special security procedures, (4) can be carried on with no special requests, (5) can be operated by anyone, (6) should have a rechargeable battery that can run for at least 10–16 hours without battery recharging or replacement, (7) should require only simple, or infrequent, calibration, (8) should allow time and date stamping for all data, and (9) easy interfacing to a computer for downloading to a central database. Testing was done onboard number of actual flights, approximately 15 flights, and the following were some of the observations on most flights that were tracked:

1. cabin altitude average remained below 1000 ft,

2. CO_2 levels ranged between 900 and 1700 ppm,

3. cabin temperature ranged between 22 and 29°C (71.6–84.2 F),

4. relative humidity ranged between 35 and 50% at the beginning of the flights and dropped to 10–25% as flight progressed, and

5. sound level was approximately 86 dBA.

Other studies investigated air fume, smoke and other related incidents on board actual flights and the associated incurred costs that can result due to delays and cancellations in corresponding flights. Shehadi et al. [40] reviewed databases from NASA, US Federal Aviation Administration (FAA), US Department of Transportation (DoT) and analyzed incidents on 33 aircraft models and sub-models. The study considered the different engines' and auxiliary power units' (APU) makes and models. The analysis showed that thousands of flights would need to be monitored to determine the root cause/source for a reported incident. Although the cost of the thought study might be high, but the associated costs due to delays and cancellations of flight due to such incidents can be approximately $32,000–$47,000 per aviation incident totaling approximately $4.5-million–$7-million US dollars in 2012 [39].

4. Conclusions

A review for studies done at the Airliner Cabin Environment Research Lab (ACERL) housed at Kansas State University was done in this chapter. Studies were categorized as per the method of study such as PIV, particle dispersion, CFD, tracer gas, and bleed air. The studies investigated airflow and turbulence characteristics in various airline cabin structures such as a generic room, Boeing 767 and 737 mockup cabins, actual aircraft, and a bleed air simulator. The objective of this chapter is to provide a database of experimental and computational study references that can be used to validate further computational investigation which in turn would help in designing a state-of-art contamination and bacterial sensors to be used on board flights and ultimately would help in improving the quality of air inside airplane cabins. The studies can also help aircraft designers improve the design of the ventilation system in the aircrafts and would provide guidelines for maintenance teams to follow better practice that can prevent previous incidents such as bleed air contamination.

Tracer gas and powder particles were used to simulate the dispersion of bacteria and viruses. It was observed, particularly with the longitudinal dispersion, that the various forms of contaminants behave in a similar dispersion manner. However, the relative bacteria concentrations appear to drop off more quickly with distance than those with the tracer gas and solid particles. This might be caused due to the fact that only viable bacteria would be counted during the studies. This might provide bias or incorrect results as some bacteria would be removed by the aircraft ventilation system and others might have become nonviable before they reach the more distant parts of the cabin. Other reasons would be the possibility that the collection plates, used to sample the bacteria, could have picked only large droplets depending on their orientation and distance. The larger droplets may settle out of the airflow before they reach the more distant parts of the cabin. Nevertheless, these data combined give a reliable quantification of the field dispersion of contaminants and provide a basis for developing or validating dispersion models.

The summarized studies give some insight into the behavior in the near (two seats or fewer from the point of release) and far fields. Evidence of large flow structures is evident in most studies. Also, there is evidence that these structures are chaotic. For example, the tracer gas data showed poor repeatability in the vicinity of the injection, but they had good repeatability at other locations. This chaotic nature makes it difficult to model and predict concentrations in the near-field region.

Tracer gas, PIV, and particle tracing are valuable experimental tools to predict the airflow characteristics and behavior. The results associated with these tools provide significant boundary conditions to validate and develop computational simulations and codes. Experimental and simulation investigation for the airflow inside aircraft passenger cabins can help in predicting the dispersion of unwanted particulates, viruses or bacteria in aircraft cabins and would help the development of appropriate decontamination methods and tools. This would help decrease the health hazardous and risks onboard these special environment compartments.

Conflict of interest

The authors do not have any conflicts of interest.

Nomenclature

ft	feet
m	meter
in.	inch
HEPA	high-efficiency particulate air
CFM	cubic feet per minute
GC/MS	gas chromatography-mass spectrometry
PIV	particle image velocimetry
CFD	computational fluid dynamics
LES	large eddy simulation
SPIV	stereoscopic particle image velocimetry
RANS	Reynolds Averaged Navier-Stokes
nm	nanometer
μm	micrometer

Author details

Maher Shehadi
Purdue University, West Lafayette, IN, USA

*Address all correspondence to: mshehadi@purdue.edu

References

[1] Shehadi M, Hosni M, Jones BW. Airflow and turbulence analysis inside a wide-body aircraft cabin mockup. Indoor and Built Environment. 2018;**27**(6):766-785. DOI: 10.1177/1420326X16689720

[2] Ebrahimi K, Zheng ZC, Hosni MH. A computational study of turbulent airflow and tracer gas diffusion in a generic aircraft cabin model. Journal of Fluids Engineering. 2013;**135**(11):1-15. DOI: 10.1115/1.4025096

[3] Lin C, Wu T, Horstman R, Lebbin P, Hosni M, Jones BW, et al. Comparison of large eddy simulation predictions with particle image velocimetry data for the airflow in a generic cabin model. In: Proceedings of the 10th International Conference on Indoor Air Quality and Climate; 4-9 September 2005; Beijing. China: Tsinghua University Press; 2005

[4] Shehadi M. Airflow distribution and turbulence analysis in the longitudinal direction of a Boeing 767 mockup cabin [Ph.D. thesis]. Manhattan, KS: Kansas State University; 2015

[5] Patel J, Jones B, Hosni M, Keshavarz A. Experimental investigation of ventilation effectiveness in an airliner cabin mockup. In: Proceedings of the ASME 2016 International Mechanical Engineering Congress and Exposition; 11-17 November 2016; Phoenix, AZ: ASME; 2016. DOI: 10.1115/IMECE2016-65341

[6] Hosni M, Jones BW. Measurement of airflow in an aircraft cabin using particle imaging velocimetry technique. In: Proceedings of the 4th International Conference on Indoor Air Quality, Ventilation and Energy Conservation in Buildings; 2-5 October 2001; Changsha, China: 2001

[7] Patel J, Jones BW, Hosni M. Experimental analysis of ventilation effectiveness and tracer gas dispersion in a Boeing-737 mockup cabin. In: Proceedings of the Second International Conference on Energy and Indoor Environment for Hot Climates; 26-27 February 2017; Doha, Qatar: ASHRAE; 2017

[8] Amiri S, Jones BW, Mohan K, Weisel C, Mann G, Roth J. Study of aldehydes, carbon monoxide, and particulate contaminants generated in bleed-air simulator. Journal of Aircraft. 2017;**54**(4):1364-1374. DOI: 10.2514/1.C034133

[9] Eckels S, Jones BW, Mann G, Mohan K, Weisel C. Aircraft recirculation filter for air-quality and incident assessment. Journal of Aircraft. 2014;**51**(1):320-326

[10] Omana M, Mann G, Jones BW, Eckels S. Persistence of bleed-air contaminants on high-efficiency particulate arrestance filters. Journal of Aircraft. 2016;**53**(5):1574-1577. DOI: 10.2514/1.C033762

[11] Jones B. The quality of air in the passenger cabin. In: Proceedings of Cabin Health 2002; Geneva: International Air Transport Association; 2002

[12] Jones B. Aircraft air quality: Facts, myths, issues, and actions. In: Proceedings of the Aircraft Interiors Conference & Exhibition; Fort Worth; 2002

[13] Hosni M, Jones BW. Application of particle image velocimetry for the measurement of the airflow characteristics in an aircraft cabin. ASHRAE Transactions. 2003;**109**(2):101-110

[14] Jones B. An air quality standard for aircraft. In: Proceedings of Air Quality in Passenger Aircraft, International

Aviation Conference; October 2003. London: Royal Aeronautical Society; 2003

[15] Lebbin PA, Hosni MH, Jones BW, Beck BT. Feasibility study of using various instrumentation for measurement of air velocity and turbulence intensity in a test room. ASHRAE Transactions. 2005;**111**(1):54-59

[16] Lin C, Wu T, Horstman R, Lebbin P, Hosni M, Jones B, et al. Comparison of large eddy simulation predictions with particle image velocimetry data for the airflow in a generic cabin model. HVAC&R Research. 2006;**12**(3, Special Issue):935-951

[17] Lebbin PA, Hosni MH, Jones BW, Beck BT, Lin CH, Horstman RH. Comparison of SPIV measurements for different test room air inlet configurations with the same inlet Reynolds number. ASHRAE Transactions. 2006;**112**(1):54-59

[18] Jones B. Review of the draft ASHRAE standard for air quality within commercial aircraft. In: Proceedings of the Annual Aviation Health Conference; November 2006; London; 2006

[19] Padilla A. Experimental analysis of particulate movement in a large eddy simulation chamber [M.S. thesis]. Manhattan, KS: Kansas State University; 2008

[20] Jones B. Review of ASHRAE standard 161-2007, air quality within commercial aircraft. In: Proceedings of the Symposium on Airliner Cabin Environment; 10-11 April 2008; Anaheim, California: ASTM; 2008

[21] Jones BW. Progress in characterizing and improving the airliner cabin environment since the 2002 National Research Council Report. In: Proceedings of the Symposium on Airliner Cabin Environment; 10-11 April 2008; Anaheim, California: ASTM; 2008

[22] Ebrahimi K, Zheng Z, Hosni M. LES and RANS simulation of turbulent airflow and tracer gas injection in a generic aircraft cabin model. In: Proceedings of the ASME 2010 Fluids Engineering Division Summer Meeting; 1-5 August 2010; Montreal, Canada: ASME; 2010. pp. 227-240. DOI: 10.1115/FEDSM-ICNMM2010-30552

[23] Shehadi M, Hosni M, Jones B. Experimental investigation of optimal particulate sensor location in an aircraft cabin. In: Proceedings of ASME 2010 3rd Joint US-European Fluids Engineering Summer Meeting and 8th International Conference on Nanochannels, Microchannels, and Minichannels; 1-5 August 2010; Montreal, Canada: ASME; 2010. pp. 849-850. DOI: 10.1115/FEDSM-ICNMM2010-30204

[24] Jones BW. Advanced models for predicting contaminants and infectious disease virus transport in the airliner cabin environment: Experimental data. In: Proceedings of the Transportation Research Board, Research on the Transmission of Disease in Airports and Aircraft, A Symposium; 17-19 September 2009; Washington, DC: The National Academies; 2010

[25] Loo S, Jones BW. Portable air quality monitor and wireless sensor network for cabin monitoring. In: Proceedings of the Sixth Triennial International Fire & Cabin Safety Research Conference; 25-28 October 2010; Atlantic City, New Jersey, USA; 2010

[26] Trupka A, Hosni M, Jones B. Tracer gas and smoke flow visualization assessment of a beverage cart on contaminant dispersion in a twin aisle aircraft cabin. In: Proceedings of the SAE AeroTech Congress; October 2011; Toulouse. SAE; 2011. DOI: 10.4271/2011-01-2692

[27] Trupka A, Hosni M, Jones BW. Tracer gas mapping of a beverage cart wake in a twin aisle aircraft cabin. In: Proceedings of the ASME 2011 International Mechanical Engineering Congress and Exposition; 11-17 November 2011; Denver, Colorado: ASME; 2011. pp. 685-694

[28] Beneke J, Jones BW, Hosni M. Fine particle dispersion in a commercial aircraft cabin. HVAC&R Research. 2011;**17**(1):107-117

[29] Korves T, Johnson D, Jones B, Watson J, Wolk D, Hwang G. Detection of respiratory viruses on air filters from aircraft. Letters in Applied Microbiology. 2011;**3**:306-312

[30] Ebrahimi K, Hosni M, Zheng Z. Computational study of turbulent airflow in a full-scale aircraft cabin mockup-part 1: Determination of boundary conditions at the outlet of air diffusers. In: Proceedings of the ASME 2013 Fluids Engineering Division Summer Meeting; 7-11 July 2013(b); Incline Village, NV: ASME; 2013. DOI: 10.1115/FEDSM2013-16564

[31] Ebrahimi K, Zheng Z, Hosni M. Simulation of the turbulent dispersion of 10 micron particles in a generic half-cabin model. Journal of Fluids Engineering. 2013;**135**(11):907-917. DOI: 10.1115/IMECE2011-64183

[32] Isukapalli S, Mazumdar S, George P, Wei B, Jones BW, Weisel C. Computational fluid dynamics modeling of transport and deposition of pesticides in an aircraft cabin. Atmospheric Sciences. 2013;**68**:198-207

[33] Anderson M, Jones BW, Hosni M. Impact of personal air outlets on person-to-person bio-effluent exposure in aircraft cabins. ASHRAE Transactions. 2013;**119**(2):1-8. DE-13-C054

[34] Korves T, Piceno Y, Torn L, Desantis T, Jones B, Hwang G. Bacterial communities in commercial aircraft high-efficiency particulate air (HEPA) filters assessed by PhyloChip analysis. Indoor Air. 2013;**23**(1):50-61. DOI: 10.1111/j.1600-0668.2012.00787.x

[35] Shehadi M, Hosni M, Jones BW. Airflow distribution in the longitudinal plan of a Boeing 767 mockup cabin. In: Proceedings of the ASME 2014 International Mechanical Engineering Congress and Exposition; 14-20 November 2014; Montreal, Canada: ASME; 2014. DOI: 10.1115/IMECE2014-40102

[36] Powell R, Jones B, Hosni M. Measurement of particle deposition rates in a commercial aircraft cabin. HVAC&R Research. 2014;**20**(7):770-779. DOI: 10.1080/10789669.2014.945851

[37] Mann G, Eckels S, Jones BW. Analysis of particulate size distribution and concentrations from simulated jet engine bleed air incidents. HVAC&R Research. 2014;**20**(7):780-789

[38] Keshavarz A, Jones BW, Hosni M, Beneke J. Experimental measurements for gaseous transport within an aircraft. Building and Environment. 2015;**89**:327-335

[39] Shehadi M, Jones BW, Hosni M. Bleed air contamination financial related costs on board commercial flights. SAE International Journal of Aerospace. 2015;**8**(2):310-316. DOI: 10.4271/2015-01-9007

[40] Shehadi M, Jones BW, Hosni M. Characterization of the frequency and nature of bleed air contamination events in commercial aircraft. Indoor Air. 2016;**26**:478-488. DOI: 10.1111/ina.12211

[41] Jones BW, Amiri S, Roth J, Hosni M. The nature of particulates in aircraft bleed air resulting from oil contamination. In: Proceedings of the 2017 ASHRAE Winter Meeting;

1 January-February 1 2017; Las Vegas, NV: ASHRAE; 2017. LV-17-C046

[42] Space D, Salgar A, Scheer D, Jones BW, Amiri S. Experimental determination of the characteristics of lubricating oil contamination in bleed air. In: Proceedings of the 2017 ASHRAE Winter Meeting; 1 January-February 1 2017; Las Vegas, NV: ASHRAE; 2017. LV-17-C046

[43] Shehadi M. Particles dispersion analysis in an 11-row Boeing 767 cabin mockup. Science and Technology for the Built Environment. 2019;**6**:1-10. DOI: 10.1080/23744731.2019.1573057

Chapter 2

Mass and Number and Its Chemical Composition Distribution of Particulate Matter in Different Microenvironments

Mahima Habil, David D. Massey and Ajay Taneja

Abstract

Adverse health effects from exposure to air pollution are a global challenge and of widespread concern. Therefore, the present study attempted to pave the way in the study of indoor air pollution by coarse and fine particulate matter and picturesque its relation with the different indoor microenvironment. A campaign study was conducted in the city of Taj "Agra," India in which different microenvironments were selected (i.e., offices, shops, and commercial centers). For each site, two different locations were chosen to examine the coarse particles (PM_{10} and $PM_{5.0}$) and fine particles ($PM_{2.5}$, $PM_{1.0}$, $PM_{0.5}$ and $PM_{0.25}$) concentrations and metal concentration of Zn, Pb, Ni, Fe, Cr, Cd, Mg and Cu in $PM_{2.5}$ and for their related health effects. The exposure factor and health risk assessment for carcinogenic effects due to heavy metal contaminants have also been calculated for adults working in different microenvironment by following the methodology prescribed by US EPA.

Keywords: indoor microenvironment, particulate matter, multivariate statistical analysis, carcinogenic and related health effects

1. Introduction

World Health Organization explains a healthy city is one that is continually creating and improving those corporal and communal environments and expanding those community assets that enable people to mutually support each other in achieving all the purposes of life and in developing to their maximum potential. While the basic tasks of the urban environment are climate change, the wasteful use of natural resources, the health impacts of airborne pollutants and exposure to vast arrays of hazardous chemicals. These problems are bound up with the modern lifestyles and at the same time require attention and action of many different parts of society. Environmental awareness to people is at rise, as more and more individuals are search for higher living standards and a better living surroundings. Thus the development of sound environmental policy requires both scientific information about the linkages between pollutant emissions and human health effects and value judgments about the importance of these effects relative to other social concerns.

Asia is one of the significant part of the world in the view of atmospheric aerosol loading because of the presence of growing economies like India, China and other developed and developing countries. Mechanization, expansion, financial growth and connected increase of energy demands have resulted in profound deterioration of urban air quality [1, 2]. The developing countries like India have shifted their economics from manufacturing toward services that involve information technologies. Growth in information technology have amplified the quantity and extended the use of equipment used in proximity to office worker due to which electronic media used for entertainment, telecommunications and data processing have become widespread in daily life [3, 4]. Typical examples are television, audio-visual recorders, stereo systems, and CPUs with their peripherals such as monitors and printers, scanners and copiers. There is growing concern about the levels of potentially harmful pollutants that may be emitted from office and other commercial centers equipment. Office equipment has been found to be a source of ozone, particulate matter, volatile organic compounds (VOCs) and semi-volatile organic compounds (SVOCs) [5]. Several studies have revealed more consistent correlation for the concentration of fine ($PM_{2.5}$) and inhalable (PM_{10}) particles with health than any other air pollutant [6, 7]. In cities, a major fraction of ultrafine aerosol particles (particle diameter, Dp <100 nm) cause adverse health effects in sensitive human beings more than larger particles due to their increased lung deposition efficiency [8]. With possible adverse health effects, the question of the chemical characterization of the fine particles released by such devices is of special importance [9]. Trace elements associated with $PM_{2.5}$ and below are nonvolatile in nature; they are less vulnerable to chemical alterations and remain in the form as they were emitted, even though they tend to undergo long range atmospheric transport [2, 10, 11]. Particulate metal components can have severe carcinogenic and toxic effects on occupants when inhaled in higher concentration. Short-term differences of atmospheric metal concentration have been observed in a day-to-day or even an hour-to-hour basis [1, 12]. Epidemiological studies show that these ultrafine particles cause more adverse health effects in sensitive human beings more than larger particles due to their increased lung deposition efficiency. Therefore the reduction of particulate and associated metal pollution to some acceptable levels is an important environmental issue. The objective of this study helps in pinpointing the integrated actions essential to reduce the particulate pollutant and eliminate the toxicological environment impacts of Indian urban environment. This can enhance the capacity of national environment within the city and which can be implemented to other towns and cities of India that can benefit health, quality of life and the economy.

2. Experimental setup

2.1 Site description

The city of Taj, Agra (27°10′N, 78°02′E) is located in the central part of northern India, about 204 km of south of Delhi in the Indian state of Uttar Pradesh. Agra is one of the most famous tourist spots of the country. The city is situated on the west bank of the river Yamuna 169 m above sea level. A part of the great northern Indian plains, Agra region experience a tropical climate. In winter the temperature ranges from 3.5 to 30.5°C and during summers which are hot and dry the temperature ranging from 32 to 48°C. The downward wind is south-southeast 29% and northeast 6% of the time in summer, and it is west-northwest 9.4% and north-northwest 11.8% of the time in winter. Agra has about 1.586 million population and the population density is about 1084 persons per square km (ORG, office of the registrar New Delhi, India: Ministry of Home Affairs, 2011). In this study real time series data for mass and number of coarse and fine particles were monitored in indoor environment of three different location (two shops, two commercial buildings and two offices) in different parts of Agra city (**Figure 1A** and **B**). Indoor PM5 was also collected from these sites for chemical characterization. The detailed description of these sites is shown in **Table 1**.

2.2 Sample collection

A short term study was conducted from September 2011 to November 2011 to determine number and mass concentration of coarse and fine particles, i.e., PM_{10}, $PM_{5.0}$, $PM_{2.5}$, $PM_{1.0}$, $PM_{0.5}$, and $PM_{0.25}$ in indoors of commercial centers, shops and offices in of Agra city. Chemical characterization in PM_5 was also carried for heavy metal detection. A total of 36 samples (i.e., 18 samples each for PM mass and

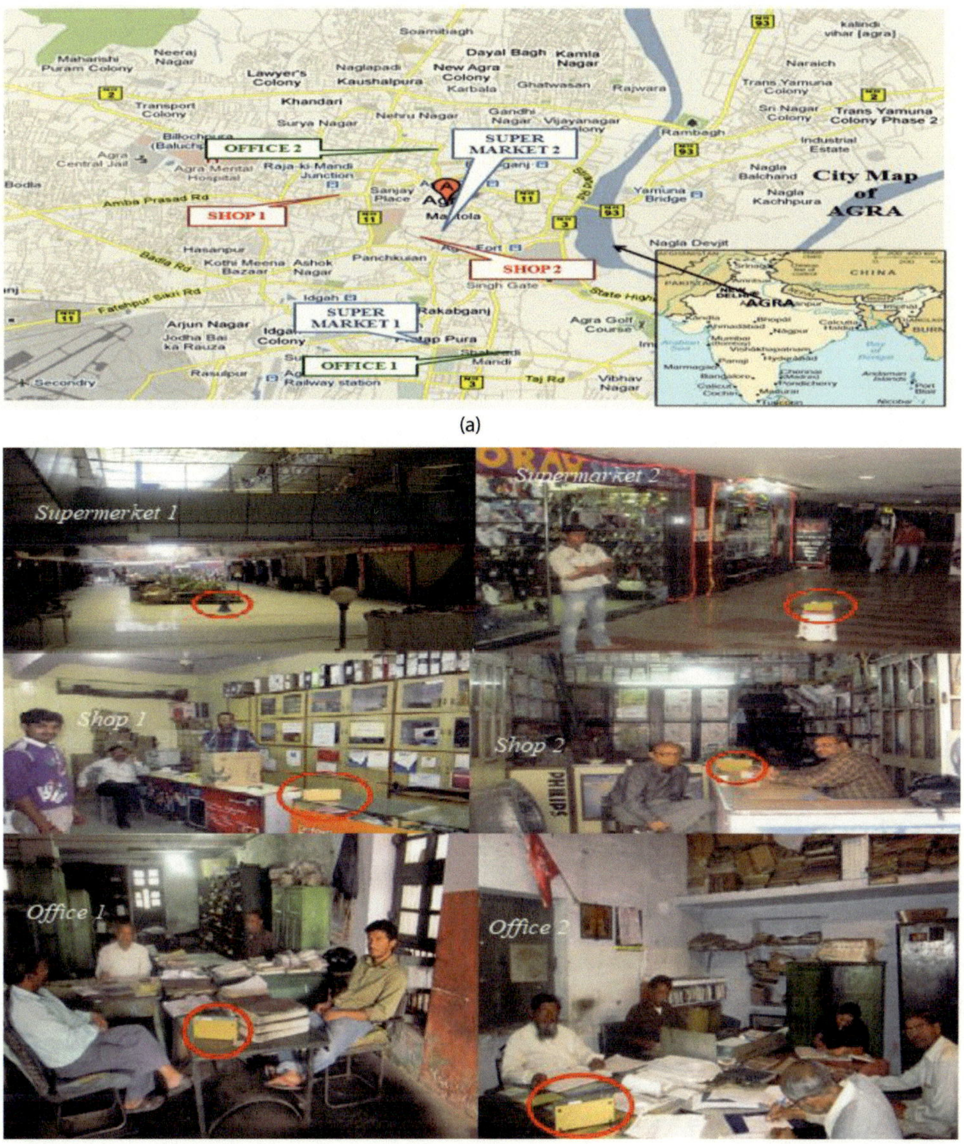

(a)

(b)

Figure 1.
(A) Map of Agra city showing different sampling sites and (B) different sampling sites.

Sampling site/type	Conditions	Building age (years)	Height	Working area (m^2)	Traffic/greenery/ trees	Ventilation
Supermarket 1	High population area, congested, made of iron, brick and cement	2	5–10	600	High traffic with no greenery around, situated in congested commercial area	Air cooling system but mostly kept close
Supermarket 2	High population area, made of iron, brick and cement	3	5–10	550	High traffic with no greenery around, situated in congested old market area	Improper ventilation system with no use of exhaust
Shop 1	Congested and very populated area of the city, made of bricks and cement	20	5	181	Heavy traffic throughout the day with no greenery around, situated in congested commercial area	Improper ventilation
Shop 2	Congested and very populated area of the city, made of bricks and cement	20	5.5	192	Heavy traffic throughout the day with no greenery around, situated in congested old market area	Improper ventilation
Office 1	Congested and very populated area of the city, made of bricks and cement	12	6	287	High traffic with less greenery, situated in congested residential area	Improper ventilation with no use of exhaust
Office 2	Congested and very populated area of the city, made of bricks and cement	40	5	250	High traffic with no greenery, situated in congested commercial area	Improper ventilation with no use of exhaust

Table 1.
Detail description of sampling sites.

number concentration) and 36 samples for chemical characterization were collected from these sites at the same time. Grimm Aerosol Spectrometer model (1.109) (**Figure 2**) was selected for the monitoring of coarse and fine particles, it runs at a flow rate of 1.2 L/min ± 5% constant with controller for continuous measurement during the sampling period. It measures mass in (μg m^{-3}) and number in (particles/m^3). The sampler measures particles from 0.25 to 32 μm range in 31 channel sizes, each unit is certified by National Institute of Standards and Technology, monodisperse latex on the size of channels calibrated by www. GRIMM Aerosols.c om. To improve the time resolution, the range was limited to 0.25–10 μm. The sampling equipment was housed such that it was as compact as possible and positioned indoors to cause minimal intrusion to the occupants. The instrument was generally positioned in the center of the rooms where people spent most of their time. Inlet heads were positioned as close as possible to head height. The instrument was set to average the data over 15 min to reduce the response time and to enable the identification of individual sources. The GRIMM particle measuring system is equipped with GRIMM 1174 Software for data acquisition.

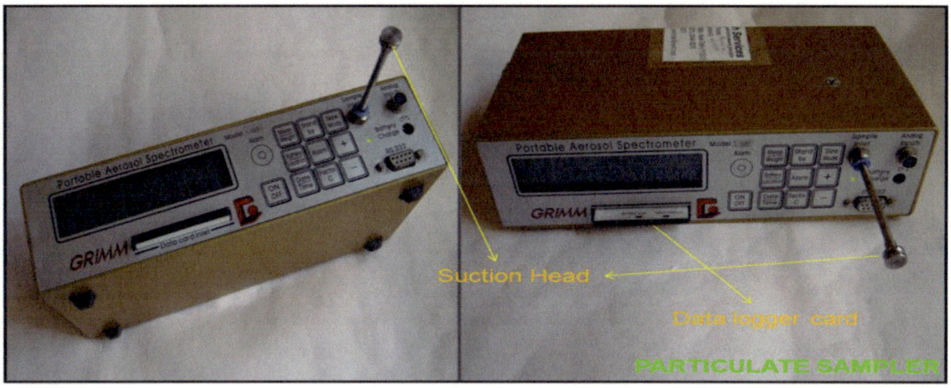

Figure 2.
GRIMM aerosol spectrometer (model: 1.109).

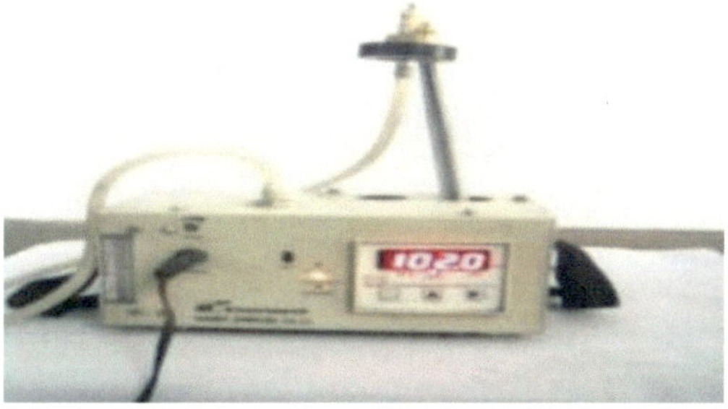

Figure 3.
Handy sampler APM 821.

After sampling the filters were thrice weighted before and after sampling using four digit (Wenser, Model No. MAB 120) with sensitivity off 2 ± mg and in the 220–20 mg range. Before the samples were equilibrated in desiccators at 20–30°C and relative humidity of 30–40% in humidity controlled room for 24 h. Filters cassettes were used to carry weight filter papers to the sampling sites, there filters were transferred to filter holders and placed on the sampling plates. Exposed filters along with the holders were then wrapped with aluminum foil, and taken back to the laboratory and placed in the desiccators. Field blanks were collected with exposed filters, they were latter weighted and were stored in refrigerator at 4°C to prevent the evaporation of volatile components [13]. Handy sampler Model No. 821 (Envirotech, New Delhi make) (**Figure 3**) was used for PM_5 chemical characterization which was maintained at a flow rate of 2 L/min and YES-IAQ monitor model No.206 (**Figure 4**) for recording the air exchange rate.

2.3 Chemical analysis

The exposed filter papers were digested in analytical grade (Merk) HNO_3 and kept on hot plate at the temperature of 4–60°C for 90 min. The solution was diluted up to 50 ml with distilled deionized water and stored in polypropylene bottles at 4°C till analysis. Analysis for metals (Fe, Br, Pb, Ba, Zn, Sb, Cu, Cd, Hg) was done on AAS (AAnalyst 100, Perkin Elmer) (**Figure 5**) present in our departmental analytical lab. The tested suites of elements were related to specific combustion sources using Principal component analysis statistical techniques.

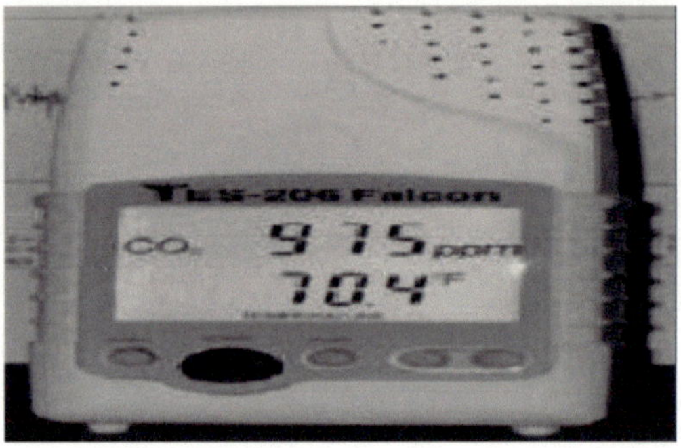

Figure 4.
Handy sampler YES-206.

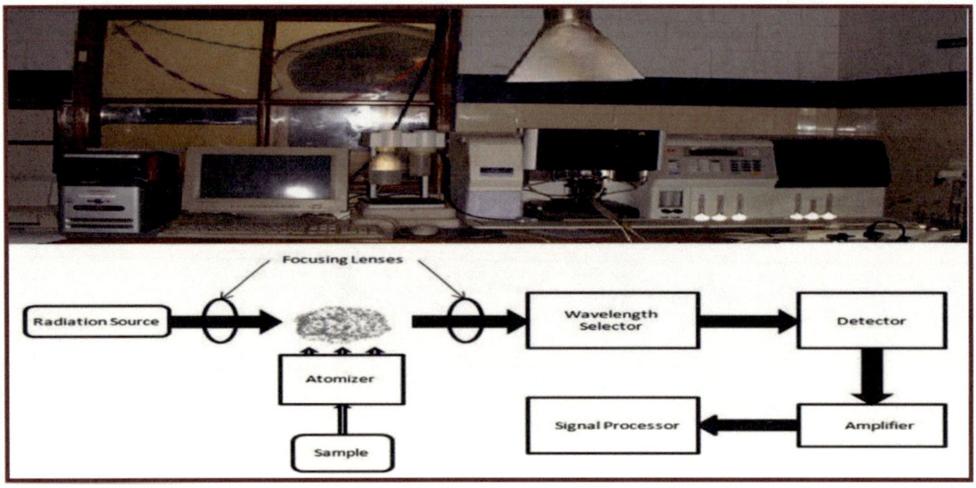

Figure 5.
Atomic absorption spectrophotometer (Perkin Elmer, AAnalyst 100), with schematic diagram.

3. Result and discussion

3.1 Particulate concentration

A total of 12 samples per month (i.e., six for mass concentration and six for number concentration) for PM_{10}, $PM_{5.0}$, $PM_{2.5}$, $PM_{1.0}$, $PM_{0.5}$, and $PM_{0.25}$ were collected from three different indoor microenvironments. **Tables 2** and **3** gives the statistical summary of particulate mass and number concentrations along with temperature, CO_2, humidity and air exchange rate during the total sampling days. During the campaign study the mean PM_{10}, $PM_{5.0}$, $PM_{2.5}$, $PM_{1.0}$, $PM_{0.5}$ and $PM_{0.25}$ mass concentration and standard deviation (SD) was 324.17 ± 46.70, 270.27 ± 42.66, 223.41 ± 48.19, 137.47 ± 23.43, 99.84 ± 20.39 and 52.34 ± 11.45 $\mu g\ m^{-3}$ at supermarket sites, 324.57 ± 47.13, 271.30 ± 40.63, 225.44 ± 49.79, 137.89 ± 23.86, 99.41 ± 20.72 and 53.07 ± 11.36 $\mu g\ m^{-3}$ at shop sites respectively and 327.00 ± 47.03, 272.98 ± 40.03, 227.44 ± 50.54, 139.17 ± 23.75, 101.33 ± 20.75 and 56.13 ± 11.58 $\mu g\ m^{-3}$ at office sites respectively and for number concentrations for coarse and fine particles, mean values of PM_{10}, $PM_{5.0}$, $PM_{2.5}$, $PM_{1.0}$, $PM_{0.5}$ and $PM_{0.25}$ was 564,050 ± 915,78.43, 320,394 ± 393,85.52, 193,678 ± 17,880.25, 174,101 ± 23,865, 158,428 ± 29,089.22 and 73,378 ± 22,638

SUPERMARKETS	PM_{10} µgm³	$PM_{5.0}$ µgm³	$PM_{2.5}$ µgm³	$PM_{1.0}$ µgm³	$PM_{0.5}$ µgm³	$PM_{0.25}$ µgm³	CO_2 PPM	TEMP °C	HUMIDITY %	AIR EXCHANGE RATE h^{-1}
AVERAGE	324.17	270.27	223.41	137.47	99.84	52.34	545.97	30.96	49.31	2.45
SD	46.70	42.66	48.19	23.43	20.39	11.45	10.78	2.01	1.03	0.37
MAXIMUM	329.00	281.11	243.27	154.55	110.54	57.79	555.49	32.45	50.39	2.86
MINIMUM	320.31	249.48	188.58	107.25	83.71	46.04	534.27	28.67	48.34	2.14
SKEWNESS	0.93	-1.73	-1.68	-1.69	-1.47	-0.63	-0.87	-1.50	0.45	1.13
SHOPS										
AVERAGE	324.57	271.30	225.44	137.89	99.41	53.07	373.37	33.86	58.76	5.24
SD	47.13	40.63	49.79	23.86	20.72	11.36	19.18	3.04	3.39	0.10
MAXIMUM	330.41	283.64	238.17	156.69	112.05	59.51	389.45	36.48	62.15	5.36
MINIMUM	319.99	251.06	192.93	106.56	82.36	46.48	352.15	30.52	55.37	5.17
SKEWNESS	0.94	-1.61	-1.61	-1.63	-1.19	-0.10	-1.12	-1.00	0.01	1.57
OFFICES										
AVERAGE	327.00	272.98	227.44	139.17	101.33	56.13	423.53	33.63	58.25	4.57
SD	47.03	40.03	50.54	23.75	20.75	11.58	8.41	2.71	3.59	0.27
MAXIMUM	333.87	283.04	245.65	157.69	113.81	62.27	432.78	36.42	62.01	4.86
MINIMUM	319.91	253.03	194.00	105.60	82.63	48.29	416.35	31.01	54.85	4.32
SKEWNESS	-0.14	-1.73	-1.71	-1.70	-1.46	-1.01	1.04	0.29	0.45	0.75

Table 2.
Statistical summary of mass concentration during the sampling duration at sampling sites.

SUPERMARKETS	PM_{10} µgm³	$PM_{5.0}$ µgm³	$PM_{2.5}$ µgm³	$PM_{1.0}$ µgm³	$PM_{0.5}$ µgm³	$PM_{0.25}$ µgm³	CO_2 PPM	TEMP °C	HUMIDITY %	AIR EXCHANGE RATE h-1
AVERAGE	564050	320394	193678	174101	158428	73378	546.89	31.15	49.31	2.48
SD	91578.34	39385.52	17880.25	23865.87	29089.22	22638.62	10.55	1.89	1.05	0.37
MAXIMUM	708054	387299	239159	229203	198148	98297	556.23	32.56	50.40	2.88
MINIMUM	385325	240431	169085	132835	97426	25012	535.45	29.00	48.30	2.15
SKEWNESS	-0.30	-0.37	1.20	0.47	-0.76	-0.66	-0.86	-1.48	0.31	0.89
SHOPS										
AVERAGE	589882	349888	206648	181495	166050	77619	364.92	33.86	58.76	5.26
SD	98489.67	39072.42	25422.77	24131.06	28853.73	22858.65	22.08	3.10	3.46	0.09
MAXIMUM	723225	417693	255400	236318	203617	105994	390.25	36.58	62.22	5.37
MINIMUM	393102	263745	170290	141152	104224	29445	349.75	30.48	55.30	5.20
SKEWNESS	-0.72	-0.49	0.57	0.30	-0.73	-0.64	1.63	-0.91	-0.01	1.64
OFFICES										
AVERAGE	622352	352319	232186	193769	178172	85121	423.59	33.65	58.31	4.64
SD	77730.91	38052.23	35323.51	28899.68	24245.03	24879.46	7.63	2.71	3.62	0.31
MAXIMUM	728091	416462	293039	250985	212192	116927	432.26	36.44	62.10	4.99
MINIMUM	479345	264499	164632	142149	127194	33026	417.85	31.01	54.90	4.38
SKEWNESS	-0.60	-0.72	-0.23	0.13	-0.70	-0.56	1.47	0.25	0.48	1.11

Table 3.
Statistical summary of Number concentration during the sampling duration at sampling sites.

particles/L at supermarket sites, 589,882 ± 98,489.67, 349,888 ± 39,072.42, 206,648 ± 25,422.77, 181,495 ± 24,131.06, 166,050 ± 28,853.73 and 77,619 ± 22,858.65 particles/L at shop sites respectively and 622,352 ± 77,730.91, 352,319 ± 38,052.23, 232,186 ± 35,323.51, 193,769 ± 28,899.68, 178,172 ± 24,245.03 and 85,121 ± 24,879.46 particles/L at office sites respectively.

On applying the one way ANOVA (SPSS 10.0) to the mean values of particulate concentrations at all the sites for each location significant values found for PM_{10}, $PM_{5.0}$, $PM_{2.5}$, $PM_{1.0}$, $PM_{0.5}$ and $PM_{0.25}$ were close to 1 or were approximately 1. For the supermarkets it varied from 0.931 to 0.997, for shops it varied from 0.942 to 0.998 and for offices they varied from 0.938 to 0.999, indicating that there is no significant difference between the concentrations of two types of similar microenvironment and thus have similar kind of sources which lead to the generation of particulate pollutant in their indoor environment. Due to the above reason, the discussion made in this report is explained on the basis of average concentration of two types of similar microenvironment rather than six places individually. The study period for mass and number concentration trend of PM_{10}, $PM_{5.0}$, $PM_{2.5}$, $PM_{1.0}$, $PM_{0.5}$ and $PM_{0.25}$ from September 2011 to November 2011 in supermarket, shops and offices are given in **Figure 6A** and **B**. On comparing with the standards

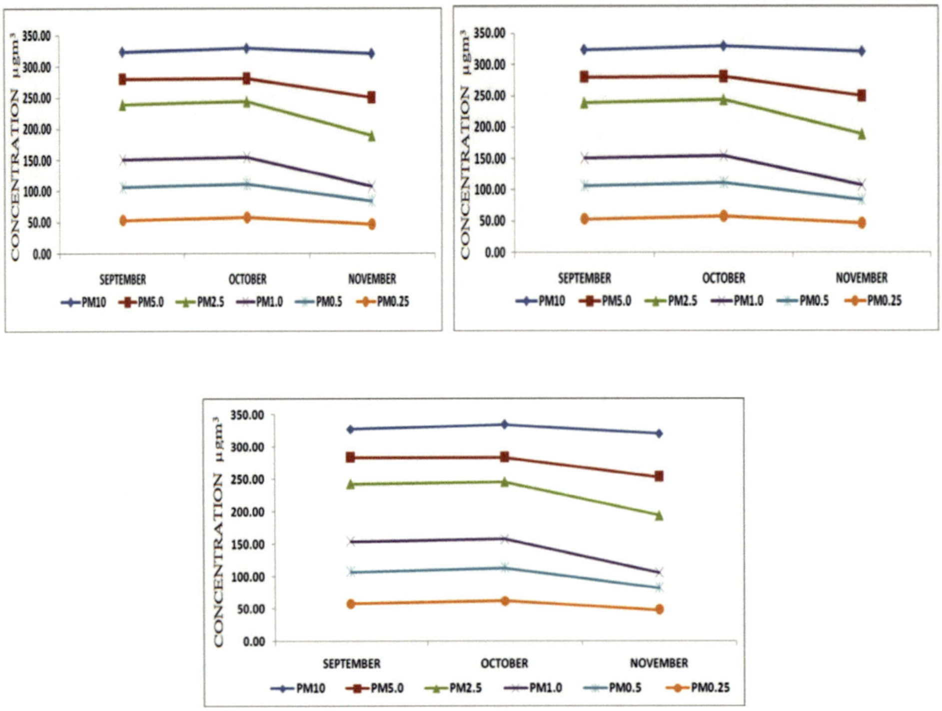

(A)

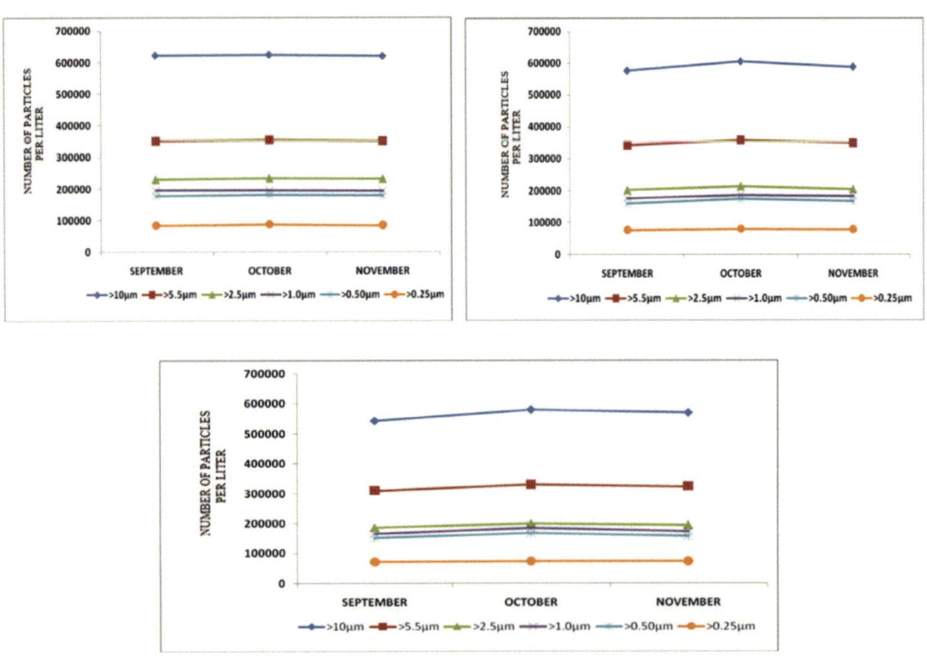

(B)

Figure 6.
(A) Mass concentration of PM_{10}, $PM_{5.0}$, $PM_{2.5}$, $PM_{1.0}$, $PM_{0.5}$ and $PM_{0.25}$ at supermarkets, shops and offices from September 2011 to November 2011, and (B) number concentration of PM_{10}, $PM_{5.0}$, $PM_{2.5}$, $PM_{1.0}$, $PM_{0.5}$ and $PM_{0.25}$ at supermarkets, shops, and offices from September 2011 to November 2011.

given by WHO guidelines (24 h mean = 25 μg m^{-3}, 50 μg m^{-3} for $PM_{2.5}$ and PM_{10}), $PM_{2.5}$ exceeded 9 times and PM_{10} exceeded 6.5 times in all the indoor microenvironment (i.e., supermarkets, shops, and offices). On comparing with NAAQS standards (24 h mean = 60 μg m^{-3}, 100 μg m^{-3} for $PM_{2.5}$ and PM_{10}), Coarse to fine particles exceeded 6 to 4 times at all the sampling locations. For coarse and fine particles similar kind of trend was obtained at all the different microenvironments. However, the mass and number concentration trends for coarse and fine particles were somewhat higher for all the particle sizes in the offices in comparison to shops and supermarkets. The higher concentration trend in the offices can be due to particle resuspension from vacuum cleaning, sweeping, low air exchange rate as or due to the movements of office workers [14, 15]. PM concentrations are also greatly affected in the offices by the use of printers and multitask devices [16]. During the campaign study a slight increase was notice in the PM concentrations during month of October in comparison to September and November. On applying the one way ANOVA (SPSS 10.0) to the mean values of particulate concentrations at all the working environment; significance values were found forPM_{10}, $PM_{5.0}$, $PM_{2.5}$, $PM_{1.0}$, $PM_{0.5}$ and $PM_{0.25}$ were close to 1 or approximately 1. They varied from 0.923 to 0.998 at two offices site, 0.918 to 0.993 at two shop sites and 0.920 to 0.987 at two commercial center sites indicating that there is no significant difference between the concentrations of same type of working environment and thus have similar kind of sources which lead to the generation of particulate pollutant in their environment.

3.2 Full day variation trend

The diurnal trend of particulate number and mass concentration during the sampling duration has been monitored continuously throughout the night and day in indoors at the supermarket, shop and office (**Figure 7A** and **B**). Full-day variation trend of particulate pollutant around the clock covers all the indoor activities taking place. The highest mass and number concentration peaks were observed basically during the morning hours from 9:00 to 10:00 AM and late in the evening hours from 18:00 to 19:00 PM. The maximum concentrations of the particulate matter during this time can be due to resuspension generated by traffic and other human activities, as these sites are mostly adjacent to busy traffic roads of the city (**Figure 1A**). As a result concentrations reaches maximum during the rush hours in the evening and morning [17]. The low concentrations for all the particles are observed during 3:00–4:00 AM in the early morning hours. Whereas, during the working hours low concentrations were reported between 14:00 and 15:00 PM in the afternoon hours at all the microenvironments. All the coarse and fine particulate sizes showed similar trends.

3.3 Inter particulate ratios

For better understanding of these particles in the different indoor environment, inter particulate ratios have also been evaluated and reported in **Table 4**. The average contribution of finer particles (i.e., $PM_{0.25}$, $PM_{0.50}$, $PM_{1.0}$, and $PM_{2.5}$) to coarse particles (i.e., $PM_{5.0}$ and PM_{10}) in indoors during the study period for September was around 44.7% at supermarket sites, 44.8% at shops sites and 47% at office sites. In October it was 45.7% at supermarket sites, 46.2% at shops sites and 47.9% at office sites. In November it was 37.1% at supermarket sites, 37.5% at shops sites and 38.0% at office sites. Particles especially $PM_{2.5}$ and below are resuspended in air with high intensive activities during indoor activities when there is low ventilation rates due to closed doors and windows, which get reduced to some level during the infiltration from higher air exchange rates [18]. The shops are more ventilated in comparison to offices and supermarkets (**Tables 2** and **3**).

At offices contribution of finer particles to coarse particles is 44% while at the shops and supermarket is around 42%. This suggests that office sites are more exposed to finer particles then in comparison to shops or supermarkets.

3.4 Metal concentrations

Characterization of PM components, including inorganic elements, is of central importance in proposing mechanisms for health effects and in source apportionment studies [19, 20]. Data obtained by chemical analysis for seven metals in $PM_{2.5}$ particulate size collected from 36 samples of different indoor environment of supermarkets, shops and offices sites is shown in **Table 5**. Observations on the basis of these tables are as follows:

(A)

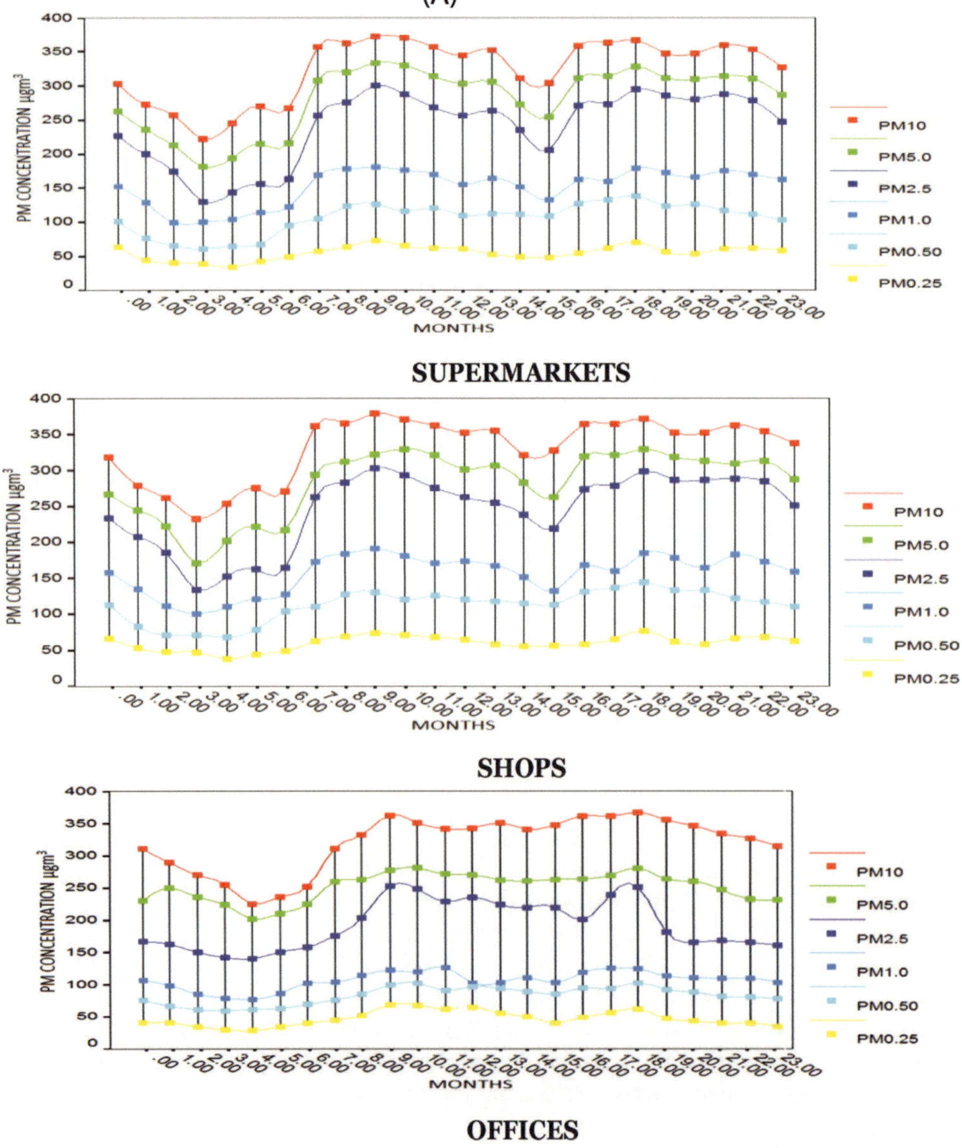

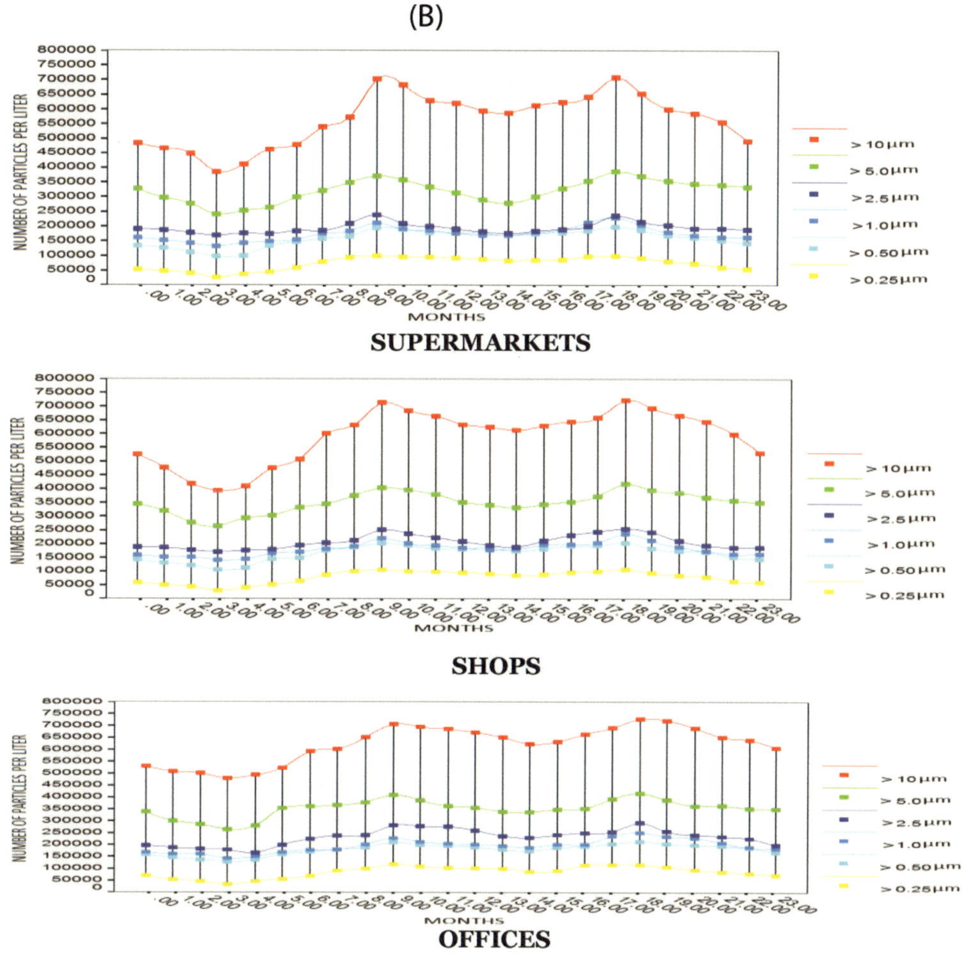

Figure 7.
(A) Full day variation in µg m^{-3} in different indoor microenvironment and (B) full day variation in particles/L in different indoor microenvironment.

SUPERMARKET	PM 5/10	PM 2.5/10	PM 1.0/10	PM 0.5/10	PM 0.25/10	PM 2.5/5	PM 1/5	PM 0.5/5	PM 0.25/5	PM 1/2.5	PM 0.5/2.5	PM 0.25/2.5	PM 0.5/1	PM 0.25/1	PM 0.25/0.5
SEPTEMBER	0.87	0.74	0.47	0.33	0.16	0.85	0.54	0.38	0.19	0.63	0.44	0.22	0.70	0.35	0.51
OCTOBER	0.85	0.74	0.47	0.34	0.18	0.87	0.55	0.39	0.21	0.64	0.45	0.24	0.72	0.37	0.52
NOVEMBER	0.78	0.59	0.33	0.26	0.14	0.76	0.43	0.34	0.18	0.57	0.44	0.24	0.78	0.43	0.55
SHOP															
SEPTEMBER	0.86	0.74	0.47	0.32	0.16	0.85	0.54	0.37	0.19	0.63	0.44	0.22	0.69	0.35	0.51
OCTOBER	0.86	0.74	0.47	0.34	0.18	0.86	0.55	0.40	0.21	0.64	0.46	0.24	0.72	0.38	0.53
NOVEMBER	0.78	0.60	0.33	0.26	0.15	0.77	0.42	0.33	0.19	0.55	0.43	0.24	0.77	0.44	0.56
OFFICE															
SEPTEMBER	0.86	0.74	0.47	0.33	0.18	0.86	0.55	0.38	0.20	0.64	0.44	0.24	0.70	0.38	0.54
OCTOBER	0.85	0.74	0.47	0.34	0.19	0.87	0.56	0.40	0.22	0.64	0.46	0.25	0.72	0.39	0.55
NOVEMBER	0.79	0.61	0.33	0.26	0.15	0.77	0.42	0.33	0.19	0.54	0.43	0.25	0.78	0.46	0.58

Table 4.
Inter Particulate Ratios at sampling sites.

3.4.1 Metal concentrate particles

At supermarket sites the sum of the average concentrations for fine particles were found to be 223.41 µg m^{-3} and ranged from 243.27 to 188.85 µg m^{-3} in indoors,

Metal concentration (μg m^{-3})	Zn	Ni	Cr	Mn	Cu	Fe	Pb
Supermarket							
Average	0.60	0.04	0.09	1.36	0.07	0.38	0.17
SD	0.02	0.02	0.01	0.13	0.01	0.04	0.03
Maximum	0.62	0.06	0.10	1.50	0.08	0.42	0.20
Minimum	0.58	0.03	0.08	1.24	0.06	0.35	0.15
Shops							
Average	0.62	0.07	0.10	1.52	0.09	0.42	0.21
SD	0.03	0.02	0.02	0.13	0.01	0.03	0.03
Maximum	0.65	0.09	0.12	1.65	0.10	0.45	0.25
Minimum	0.60	0.06	0.09	1.40	0.08	0.40	0.19
Offices							
Average	0.67	0.08	0.12	1.78	0.13	0.45	0.25
SD	0.03	0.02	0.03	0.15	0.02	0.03	0.04
Maximum	0.70	0.10	0.15	1.95	0.15	0.48	0.29
Minimum	0.65	0.07	0.10	1.66	0.11	0.42	0.22

Table 5.
Statistical profile of metal concentrations in $PM_{2.5}$ (N = 36).

at shops site the concentration was 225.44 μg m^{-3} and ranged from 238.17–192.93 μg m^{-3} in indoors and whereas at offices sites the concentration was found to be 227.44 μg m^{-3} and the range from 245.65 to 194.00 μg m^{-3} in indoor environment respectively. The total analyzed parameters contributed 65% at supermarket sites, 70% at shop sites while 75% at the offices site of the particulate concentration respectively. The trends in increasing order of metal concentrations at supermarket, shop and office sites are as follows:

$$\mathrm{Ni} < \mathrm{Cu} < \mathrm{Cr} < \mathrm{Pb} < \mathrm{Fe} < \mathrm{Zn} < \mathrm{Mn} \quad (1)$$

A similar kind of trends were found for metal concentrations in all the three microenvironment, indicating one or more similar kind of sources contributing to these microenvironment, being present in similar kind of commercial areas of the city. Out of 0.14% contribution of analyzed metals in PM_5 at supermarket percentage contribution of each metal is shown in **Figure 8A** while from the total

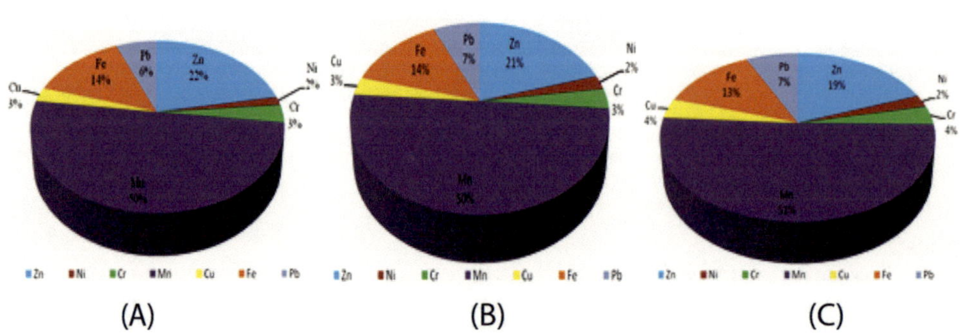

Figure 8.
(A–C) Percent contribution of each metal at supermarket, shop, and office sites.

contribution of 0.16% at shop sites and 0.20% at office sites percentage contribution of each metal is shown in **Figure 8B** and **C**. Global emissions reported shows that natural and anthropogenic sources can contribute to the principal aerosol classes, but values change according the local scenario (coarse and fine) of atmospheric particulate matter (PM). About 10–20% of the aerosols can be characterized as anthropogenic on a global scale [21], but these values may drastically change due to local scenarios, human activities, and the prevailing particle cutoff. The principal component analysis is the most common multivariate statistical methods applied in environmental studies. The SPSS software package version 10.0 was used for the multivariate analysis. The levels of various elements vary by different orders of magnitude and hence the principal component analysis was applied to the correlation matrix.

3.5 Multivariate principal component analysis

Principal component analysis (PCA) is a well-established tool for analyzing structure in multivariate data sets [22]. A varimax rotated factor analysis was performed to identify the major sources responsible for the particulate pollutants and the sampling sites. It is a statistical method; in which a set of multiple intercorrelated variables are replaced by small number of independent variables or factors by orthogonal transformations also called as rotations. This is achieved by analyzing the correlation matrix of the variable, i.e., by computing their Eigen values and the Eigen vectors. "Factor loadings" obtained after the varimax rotation give the correlation between the variables and the factors. Data which is included in the matrix only if their Eigen values of tat factor is bigger than 1. The varimax technique was adopted for rotation of the factor matrix to allocate the initial matrix into one that was easier to understand. For this SPSS version 10.0 was used to perform factor analysis. At supermarket, shop and offices sites we have mostly multistory modern types of construction, as they have been built recently; usually their outside environment is of high traffic during the day with both heavy and light motor vehicles. These are in the busy commercial places of the city with many kinds of other activities like major hospitals, hotels, railway station, big or small restaurants and banks etc.

3.5.1 Supermarket

At supermarket sites indoors three sources identifying 90% of metal concentration were identified (**Table 6**). Zn, Ni, Cr and Mn represent the factor 1 with 36% variance. The common source attributed to the smoking of the occupants in the indoor environment [23]. Factor two is represented by Cu and Fe with 31% variance and was attributed to resuspension of dust due to indoor human activities [23]. Third factor is comprised of Pb with 24% and is attributed to emission of paints from wall, ceiling and furniture [24].

3.5.2 Shops

At shop sites indoors three sources identifying 93% of metal concentration were identified (**Table 7**). Zn, Cr and Mn represent the factor 1 with 37% variance. The common source attributed to the smoking of the occupants in the indoor environment [25]. Factor two is represented by Ni, Cu and Fe with 28% variance

Rotated component matrix at office sites			
	1	**2**	**3**
Zn	**0.71**	0.29	0.43
Ni	**0.78**	0.44	0.30
Cr	**0.74**	0.43	0.19
Mn	**0.79**	0.29	0.48
Cu	0.23	**0.75**	0.22
Fe	0.38	**0.62**	0.45
Pb	0.44	0.31	**0.76**
Total	2.51	2.16	2.14
% of variance	35.88	30.79	23.62
Cumulative %	35.88	66.67	82.29

Table 6.
Factor analysis at supermarket sites.

Rotated component matrix at office sites			
	1	**2**	**3**
Zn	**0.62**	0.47	0.46
Ni	0.48	**0.68**	0.42
Cr	**0.70**	0.33	0.26
Mn	**0.83**	0.34	0.41
Cu	0.44	**0.89**	0.38
Fe	0.39	**0.76**	0.30
Pb	0.42	0.34	**0.87**
Total	2.89	1.99	1.89
% of variance	37.28	28.45	26.99
Cumulative %	37.28	69.73	96.72

Table 7.
Factor analysis at shop sites.

and was attributed to electrical wiring or appliances [1]. Third factor is comprised of Pb with 27% and is attributed to emission of paints from wall, ceiling and furniture [24].

3.5.3 Offices

At office sites indoors three sources identifying 92% of metal concentration were identified (**Table 8**). Zn, Ni, Cr and Mn represent the factor 1 with 30% variance. The common source attributed to the smoking of the occupants in the indoor environment [25]. Factor two is represented by Cu and Fe with 32% variance and was attributed to resuspension of dust due to indoor human activities [23]. Third factor is comprised of Pb with 30% and is attributed to emission of paints from wall, ceiling, and furniture [24].

Rotated component matrix at office sites			
	1	**2**	**3**
Zn	**0.57**	0.47	0.46
Ni	**0.55**	0.38	0.47
Cr	**0.54**	0.62	0.48
Mn	**0.62**	0.47	0.32
Cu	0.41	**0.82**	0.40
Fe	0.39	**0.84**	0.34
Pb	0.32	0.47	**0.87**
Total	2.40	2.24	2.10
% of variance	30.22	32.03	29.99
Cumulative %	30.22	66.25	96.24

Table 8.
Factor analysis at office sites.

3.6 Risk assessment from carcinogenic metals in different working environment

Excess cancer risks (ECRs) were calculated using the unit risk and the PM-bound concentration of the metals which represents the total concentration of the metals. ECRs can be calculated simply by using the formula given below [26, 27]:

$$\text{Excess cancer risk (inhalation)} = \text{concentration of pollutant}\ (\mu g\ m^{-3}) \times \text{unit risk}\ (\mu g\ m^{-3})^{-1} \tag{2}$$

The information on the carcinogenic types and the unit risks of the metals was obtained from the US EPA database for IRIS (Integrated Risk Information System) [28]. Fine particles, $PM_{2.5}$-bound metals such as Cd, Cr, Ni and Pb are the known carcinogenic metals which can cause serious health risks to occupants in these microenvironments. They are introduced to the occupants by exposure through the inhalation pathway [27]. Cadmium have been categorized as the B1 carcinogen, while Cr (VI) is classified as group A which indicates that it is a known human carcinogen by the inhalation route of exposure. Therefore, the concentration of Cr (VI) used for the carcinogenic risk assessment was calculated as one seventh of the total Cr concentration. Nickel also has been classified as group A materials, known human carcinogens. From the research findings it is evident that Ni was mainly emitted from tobacco or cigarette smoke and outdoor sources like the industrial or refinery emissions can contribute to it in an indoor environment [29]. Lead on the other hand is also a probable human carcinogen (group B2), but human evidence is inadequate and its unit risk is currently being amended by the US EPA.

$PM_{2.5}$-bound metals such as Cd, Cr and Ni, respectively, based on PM concentrations. The particles whose diameters are less than 4 mm can penetrate into the trachea, bronchi and the alveoli [28, 29]. In **Table 9**, the estimated ECR of $PM_{2.5}$-bound carcinogenic elements in the indoor environment for the average values and the 95th percentile values of Cd, Cr and Ni. The total ECRs based on the average values of Cd, Cr (VI) and Ni in $PM_{2.5}$ in different indoor environment varied from 0.47 to 0.32 $\times$ 10^{-6}, respectively. Ni had the highest

Elements	Concentration (ng m^{-3})		Inhalation unit risk ($\mu g\ m^{-3})^{-1}$	Excess cancer risk ($\mu g\ m^{-3})^{-1}$	
Offices	Average	95th percentile		Average	95th percentile
Cd	0.08	1.05	1.8×10^{-3}	0.14	2.51
Cr $^{(VI)\ (=\Sigma Cr/7)}$	0.11	2.01	1.2×10^{-2}	0.13	2.43
Ni	0.07	1.03	2.4×10^{-4}	0.16	2.76
Shops	Average	95th percentile		Average	95th percentile
Cd	0.08	1.04	1.8×10^{-3}	0.14	2.40
Cr $^{(VI)\ (=\Sigma Cr/7)}$	0.10	1.95	1.2×10^{-2}	0.12	2.37
Ni	0.06	1.02	2.4×10^{-4}	0.14	2.58
Commercial centers	Average	95th percentile		Average	95th percentile
Cd	0.05	1.01	1.8×10^{-3}	0.09	1.97
Cr $^{(VI)\ (=\Sigma Cr/7)}$	0.09	1.82	1.2×10^{-2}	0.10	2.20
Ni	0.04	1.00	2.4×10^{-4}	0.09	2.36

Table 9.
Excess cancer risks of carcinogenic elements in $PM_{2.5}$ in different indoor environment.

ECRs followed by Cd and Cr in the different indoor environment. The results indicates that occupant exposure to toxic trace metals in indoor environments can easily get cancer in different indoor working environments. Thus, trace metals in airborne fine and ultrafine particles be used in order to more accurately assess environmental and health risks. Thus chemical speciation of metal is important and should become a routine analysis in future study of air pollution.

4. Conclusion

Indoor air pollution in developing countries is recognized as a major source of health risk to the exposed population, thus there is a need to address the issue of particles, especially fine particles and their related toxicity in different indoor microenvironments. On comparing with the standards given by WHO and national standards, PM_{10} and $PM_{2.5}$ concentrations exceeded many folds at all the sampling sites. These high values indicate the need to find strategies to control particulate pollution. Metal concentrations were determined using the positive matrix factorization in $PM_{2.5}$ in different indoor working environments. The sources responsible for $PM_{2.5}$ emissions are smoking, incense burning, anthropogenic activities and use of different mechanical and electrical apparatus like computers, printers and photocopiers, etc. in the work environment. Risk assessment related to particulate pollutant was evaluated on the basis of metal contamination. The values summarized in this study represent initial estimates of emissions and their implications, which can be a useful addition to the existing literature, in particularly for the developing countries like India; where such measurements are yet under represented.

Acknowledgements

Support is duly acknowledged by Council of Science and Industrial Research (Project No. 8/109 (0010)/2011-EMR-I). I would like to thank Principal and Head of the Chemistry Department, St. John's College Agra for providing necessary facilities.

Author details

Mahima Habil[1*], David D. Massey[1] and Ajay Taneja[2]

1 Department of Chemistry, St John's College, Agra, India

2 Department of Chemistry, Dr. B.R. Ambedkar University, Agra, India

*Address all correspondence to: mahi.habil@gmail.com

References

[1] Hail M, Massey D, Tanja A. Particles in different indoor microenvironments-its implications on occupants. Building and Environment. 2016;**106**:237-244. DOI: 10.2016/j.buildenv.2016.06.036

[2] Massey D, Kul Shrestha A, Maish J, Tanja A. Seasonal trends of PM_{10}, $PM_{5.0}$, $PM_{2.5}$ & $PM_{1.0}$ in indoor and outdoor environments of residential homes located in north-Central India. Building and Environment. 2012;**47**:223-231. DOI: 10.2016/j.buildenv.2011.07.018

[3] New Burger EC, 2001. Home Computers and Internet Use in the United States, Special Studies. Washington, DC: US Census Bureau; 2000

[4] Massey D, Habil M, Taneja A. Characterization of coarse and fine particles in different microenvironments—Its implications on occupants. Advances in Environmental Biology. 2014;**8**(15): 61-66

[5] Wolkoff P, Wilkins CK, Clausen PA, Larsen K. Comparison of volatile organic compounds from processed paper and toners from office copiers and printers: Methods, emission rates, and modeled concentrations, indoor air. 1993;**3**:113-123

[6] Schwartz J. Why are people dying on higher air pollution days. Environmental Research. 1994;**64**:26-35

[7] Schwartz J, Dockery DW, Neas LM. Is daily mortality associated specifically with fine particles. Journal of Air Waste Management Association. 1996;**46**: 927-939

[8] Wehner B. Aerosol number size distributions within the exhaust plume of a diesel and a gasoline passenger car under on-road conditions and determination of emission factors. Atmospheric Environment. 2009;**43**: 1235-1245

[9] BfR—Federal Institute for Risk Assessment. Gesundheitliche Bewertung Nr. 014/2008 voL 31, März. Press release

[10] Morawska L, Zhang J. Combustion sources of particles. 1. Health relevance and source signatures. Chemosphere. 2002;**49**:1045-1058

[11] Niu JJ, Ramussen PE, Wheeler A, Williams R, Chenier M. Evaluation of airborne particulate matter and metals data in personal, indoor and outdoor environments using ED-XRF and ICP-MS and co-located duplicate samples. Atmospheric Environment. 2010;**44**: 235-245

[12] Gharaibeh AA, El-Rjoob AWO, Harb MK. Determination of selected heavy metals in air samples from the northern part of Jordan. Environmental Monitoring and Assessment. 2009;**160** (1–4):425-929

[13] USEPA. Compendium of Methods for the Determination of Inorganic Compounds in Ambient Air. Cincinnati, OH: Center for Environmental Research Information Office of Research and Development U.S. Environmental Protection Agency; 1999. p. 728

[14] Corsi RL, Siegel JA, Chiang C. Particle resuspension during the use of vacuum cleaners on residential carpet. Journal of Occupational and Environmental Hygiene. 2008;**5**:232-238

[15] Kemp PC, Dingle P, Neumeister HG. Particulate matter intervention study: A causal factor of building-related symptoms in an older building. Indoor Air. 1998;**8**:153-171

[16] Massey D, Taneja A. Emission and formation of fine particles from

hardcopy devices: The cause of indoor air pollution. In: Monitoring, Control and Effects of Air Pollution. In-Tech Publication; 2011. ISBN 978-953-307-204-3

[17] Harrison RM, Deacon AR, Jones MR, RS A. Sources and processes affecting concentrations of PM10 and PM2.5 particulate matter in Birmingham (UK). Atmospheric Environment. 1997; **31**:4103-4117

[18] Baek SO, Kim YS, Perry R. Indoor air quality in homes, office and restaurant in Korea urban areas—Indoor/outdoor relationship. Atmospheric Environment. 1997;**31**: 529-544

[19] Butler OT, Cook JM, Davidson CM, Harrington, Miles DL. Atomic spectrometry update. Environmental analysis. Journal of Analytical Atomic Spectrometry. 2009;**24**:133-177

[20] Kulshrestha A, Satsangi PG, Masih J, Taneja A. Metal concentration of $PM_{2.5}$ and PM_{10} particles and seasonal variations in urban and rural environment of Agra, India. Science of the Total Environment. 2009;**407**: 6196-6204

[21] Chung SH, Seinfeld JH. Global distribution and climate forcing of carbonaceous aerosols. Journal of Geophysical Research. 2002;**107**: 509-512

[22] Derwent RG, Middelton DR, Field RA, Goldstone ME, Lester JN, Perry R. Analysis & interpretation of air quality data from an urban roadside locations in Central London over the period from July 1991 to July 1992. Atmospheric Environment. 1995;**29**:923-946

[23] Gemenetzis P, Moussas P, Arditsoglou A, Samara C. Mass concentration and elemental composition of indoor $PM2_{.5}$ and PM_{10} in university rooms in Thessaloniki, northern Greece. Atmospheric Environment. 2006;**40**:3195-3206

[24] Jonathan MS, John DS. Indoor Environments and Health: Moving into the 21st Century. Vol. 93. American Public Health Association; 2003. pp. 1489-1493

[25] Eštoková A, Stevulová N, Kubincová L. Particulate matter investigation in indoor environment. Global NEST Journal. 2010;**12**:20-26

[26] Park EJ, Kim DS, Park K. Monitoring of ambient particles and heavy metals in a residential area of Seoul, Korea. Environment Monitoring and Assessment. 2008;**137**:441-449

[27] Hieu NT, Lee BY. Characterization of particulate matter and metals in the ambient air from a residential area in the largest industrial city in Korea. Atmospheric Research. 2010;**98**:526-537

[28] USEPA. Reference Dose (RfD): Description and Use in Health Risk Assessments. Background Document 1A, Integrated Risk Information System (IRIS). 1993

[29] Izhar S, Anubha G, Chakraborty A, Tarun G. Annual trends in occurrence of submicron particles in ambient air and health risk posed by particle bound metals. Chemosphere. 2016;**146**:582-590

Section 2

Indoor Health and Safety

Chapter 3

Workplace Health and Its Impact on Human Capital: Seven Key Performance Indicators of Workplace Health

Young Lee

Abstract

Health, a state of complete physical, mental and social well-being according to the World Health Organization, is a critical issue in the workplace as it is directly related to human capital, the most important and expensive asset of an organization. When it comes to workplace health, there are seven key performance indicators to consider. These include physical fitness, physical comfort, physical nourishment, cognitive well-being, social well-being, emotional well-being, and environmental well-being. Various environmental attributes in these seven KPIs in the workplace affect not only health but also performance and engagement of employees via their physical, mental, and social interactions within the environment. For instance, ergonomics, acoustics, lighting, thermal comfort, and olfactory comfort address the overall physical comfort while biophilic components contribute to employee cognitive functions as well as their capacity to cope with mental stress and fatigue. These seven KPIs of workplace health ultimately contribute to five positive organizational outcomes, including healthy organizational culture, higher productivity, improved individual health and safety, financial savings, and enhanced reputation of the organization. This chapter discusses critical health factors in the workplace and their contributions to the capacity of human capital at the individual as well as organizational levels.

Keywords: KPIs of workplace health, PROWELL, human performance, physical fatigue, mental restoration, organizational health outcomes

1. Global health epidemics

We now often hear that chronic diseases are the great health epidemics of our times. This expresses critical concerns about rapidly growing and spreading chronic diseases across the world. Chronic diseases have been blamed for over 60% of all deaths and are expected to grow to nearly 75% by 2020 [1]. The main driver behind this trend is a shift from physical labor-intensive primary industries to knowledge-intensive industries and a technologically advanced lifestyle, which led to physical inactivity, sedentary lifestyle, and unhealthy dietary behaviors. These factors are the causes of intermediate risks such as overweight and obesity, which explain incidents of the majority of chronic diseases such as heart disease, stroke, and diabetes.

The World Health Organization (WHO) statistics reveals that obesity across the world has nearly tripled since 1975 with 39% of adults being overweight, body mass index (BMI) 25 and over, while nearly 13% obese, BMI 30 and over, worldwide [2]. The urgency of these health issues is even prevalent in the US where nearly 70% of adults are overweight and 35% obese based on the WHO statistics. The criticality of the problems with overweight and obesity lie in their links to many illnesses including asthma, musculoskeletal disorders, cardiovascular disease, stroke, type 2 diabetes, osteoarthritis, and certain cancers [3]. Along with physical health issues, mental health issues have also increased due to financial instability, higher demands at job, work-life imbalance, and social isolation. The WHO speculates that a quarter of the world populations experience some sort of mental health issues in their life time with depression being the fourth leading cause of the disease worldwide [4].

With globally prevalent health epidemics of chronic illnesses, various disciplines have joined forces to seek and implement interventions beyond the medical and public health realms. In addition, multi-disciplinary approaches are sought as more desirable in tackling such a complex problem in a collective, systematic way, bringing professionals together in environmental design, food science, health science, neuroscience, and policy. A recent action plan from the WHO to increase physical activity clearly exhibits this approach by calling for strategies in four areas: social culture, physical environments, people's activities, and systems support [5].

The role of physical environment is gaining more traction than ever before in terms of increasing physical activities, changing behaviors, and controlling environmental toxins and pollutants for occupant's health. The sustainable building communities in particular are emphasizing occupant health to mitigate negative environmental impacts on human health. More stringent systems for healthy materials have been developed such as Health Product Declaration and Living Building Challenge Red List. Health and well-being building certification systems have emerged such as Facility Innovations toward Wellness Environment Leadership (FITWEL)® and WELL Building Standard™. Various green building rating systems such as LEED™, BREEAM® and Nabers are evolving to include more occupant health-related criteria to respond to these global epidemics.

2. Workplace health

As global concerns on health epidemics of chronic illnesses prevail, workplaces have become the center of attention in two regards: (1) the relationship between employee health and astronomical costs associated with it, and (2) a unique role of workplaces being a main source of health issues as well as a central place to implement health interventions more effectively.

2.1 Workplace health and its costs

Healthy employees are an important issue as an organization not only for higher productivity and economic benefits but also for organizational culture. People are the most important asset, human capital, to any organization. The majority of the costs related to doing business are attributed to people costs including salaries and benefits. Thus, means to support their optimal performance are important to organizations. Health in the workplace has become critical as organizations have learned that health conditions significantly affect the performance of their employees.

Employee health affects an organization in various ways, sometimes as direct costs such as healthcare costs and workers compensation claims but other times as indirect costs such as productivity, absenteeism, presenteeism, and engagement.

Healthcare costs have been the most troublesome concern in the US workplaces due to a huge amount of healthcare costs imposed on employers. The concerns with rising healthcare costs in the workplaces are also growing in Europe. And even bigger concerns exist with humongous indirect costs associated with unhealthy workforce: higher absenteeism and presenteeism, decreased productivity, and lower engagement.

In the EU, adult obesity accounts for 6% of direct healthcare costs and 12% of indirect costs [6]. Among these indirect costs are sickness absenteeism, being absent at work due to illness. There have been abundant studies on sickness absence, especially, in the public health realm for the last couple decades due to a clearer and simpler methodological advantage. In a study, sickness absence was speculated to cost UK employers over £600 per employee [7]. In more recent studies, the intangible but more impactful costs associated with presenteeism, being present at work but not productive, has become at the center of attention. A conservative calculation of the cost of presenteeism tells us that presenteeism may be nearly twice as high as absenteeism [8].

Productivity and engagement are also indirect costs frequently associated with health issues in the workplace. Productivity loss due to mental health issues explains 3–4% of GDP in the EU [9], and Gallup surveys show a close relationship between unhealthy workforce and low engagement as well as a positive relationship between engagement level and productivity level [10]. While it is important to understand the impact of health in the workplace on absenteeism, presenteeism, productivity, and engagement, we must be cautious in interpreting intangible indirect costs associated with workplace health to actual dollar amounts. This is because of discrepancies observed among the studies as well as credibility of assumptions and reliability of data used for calculations of these impacts in dollar amounts.

2.2 Workplace as a source of health issues and a place to cure them

Workplaces are known as a main source of many health issues in the contemporary society, including sedentary workstyle, work-life imbalance, and increased job demands. People spend a majority of their time at work, allegedly 2/3 of their waking time [11]. Due to the contemporary work settings prevalent in knowledge-intensive industries, sedentary work settings have become the main workplace environment where people tend to have a sedentary posture for a prolonged period. The sedentary work settings have shown an association with overweight and obesity as people who spent longer working hours in the workplace exhibited a higher body mass index [12]. In addition to physical health, the workplace is also a major source of mental health issues. Stress, depression, and anxiety were the number one cause of all absences at work and accounted for nearly half of all reported work-related illnesses [13]. Stress, especially, was the second highest among all reported work-related health issues [14].

While workplaces are a main source of health issues, workplaces also hold an advantageous situation in tackling people's health in a large group setting rather than working with individuals. Workplaces have been sought as a practical place to implement health interventions by health professions to change the prevalent health epidemics for the past several decades. This is because the workplace is where people spend a substantial amount of their time, so it naturally becomes a central place to not only easily implement health programs in social settings with necessary social supports but also reach larger groups of people to effectively control and prevent the health epidemics.

Addressing health issues in the workplace has been a long battle in the public health domain over several decades, especially, for chronic disease prevention and health promotion. Leading health organizations, including the WHO, the European Network for Workplace Health Promotion (ENWHP), and the US Centers for Disease Control and Prevention (CDC), have introduced a concept of workplace health promotion or health-promoting workplaces more than a decade ago targeting four areas: physical exercise, healthy diet, mental health, and healthy lifestyle in the workplace. Since then, various workplace health promotion programs have been suggested, implemented, and shown positive outcomes.

3. Environmental health in the workplace

There are currently growing efforts to integrate environmental health issues into the realm of public health as a comprehensive approach to tackle workplace health. Health in the physical environment of workplaces has two approaches to workplace health: environmental design to promote physical movement, healthy diet, mental health, and healthy lifestyle at work; and environmental control and monitoring of harmful toxins, pollutants, and irritants as well as provision of human comfort factors. Health-promoting environmental design focuses on environmental interventions such as spatial layout for physical movement, stair use, fitness spaces, spatial features for conscious eating, or biophilic design that integrates nature into the workspaces. Environmental control and monitoring emphasizes proper levels of indoor air quality, comfort factors in temperature, lighting, and acoustics, as well as managing toxic materials and substances in building components.

This section describes both approaches in seven topics called seven key performance indicators (KPIs) of workplace health (7 KPIs of WH). The indoor environmental features, conditions, and design approaches described in this section are derived from various theories and practices, such as biophilic design, environmental preference theory, evolutionary psychological design, and active design, as well as standards and guidelines in the built environment, such as FITWEL®, WELL™, LEED™, BREEAM® and health-related recommendations from the WHO and other relevant governmental agencies. It focuses on the major KPIs of Workplace Health in a comprehensive manner from a design and planning purpose so, further specific issues and items should be discussed in details with experts in each respective field when applying it in practice.

3.1 Seven key performance indicators of workplace health

Workplace health relevant to the environment can be addressed in seven dimensions within three health domains defined by the WHO: physical, mental, and social domains. Since health is defined as a state of complete physical, mental and social well-being [15], workplace health can be categorized into seven dimensions of well-being states: physical fitness, physical comfort, physical nourishment, cognitive well-being, social well-being, emotional well-being, and environmental well-being. These are essentially seven key performance indicators (KPIs) of workplace health (7 KPIs of WH) when it comes to the environment. Physical fitness, physical comfort, physical nourishment, and environmental well-being fall under the physical domain; cognitive well-being and emotional well-being under the mental domain; and social well-being under the social domain of health (**Figure 1**). This model is called PROWELL. The list of 7 KPIs in the PROWELL model delineates a comprehensive approach to workplace health to accomplish the overall health promotion in the workplace as various factors influence one another in the complex layers of workplace health. The 7 KPIs of WH contribute to overall five positive organizational outcomes: healthy organizational culture, higher employee productivity,

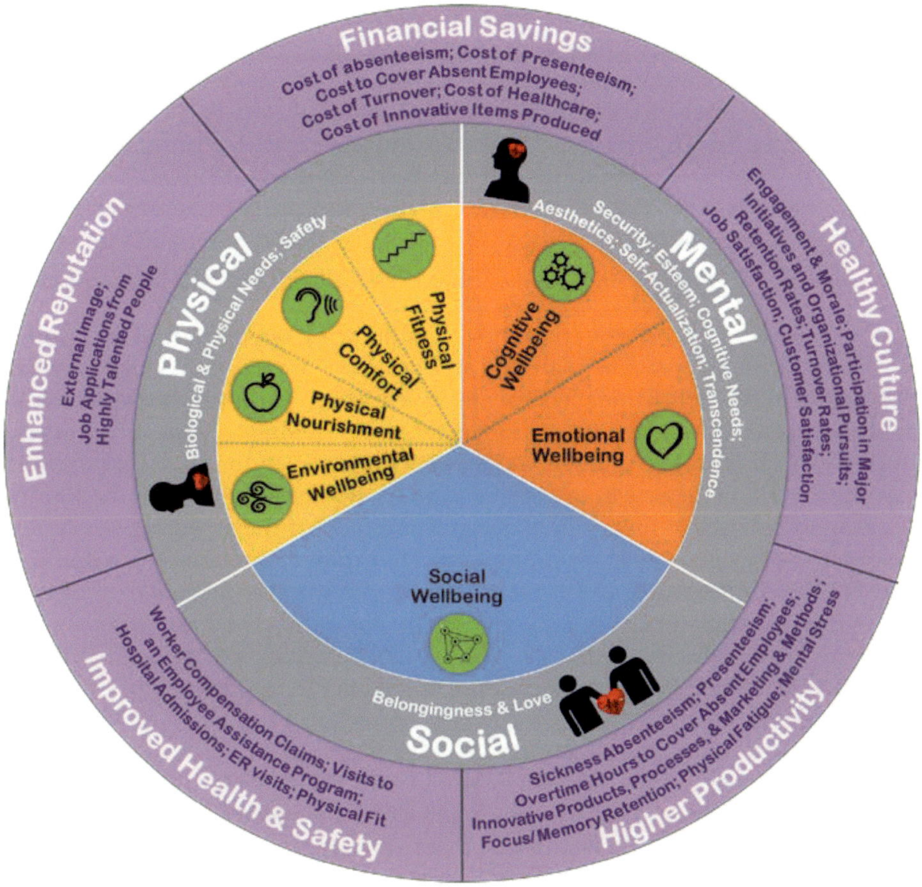

Figure 1.
Seven KPIs of workplace health in three domains and their benefits in the PROWELL model.

improved individual health and safety, enhanced company reputation, and financial savings. Health-enhanced workplaces by the 7 KPIs of WH contribute to healthy organizational culture by promoting engagement and morale, increasing retention rates, decreasing turnover rates, and increasing job satisfaction; higher productivity by reducing sickness absenteeism and presenteeism, decreasing overtime hours to cover absent employees, producing more innovative outcomes, enhancing focus and memory retention, and decreasing mental stress as well as physical fatigue; improved health and safety by reducing workers compensation claims, lower visits to an employee assistance program as well as hospital admissions and emergency visits, and better physical fit; enhanced company reputation by improved external image and increased job applications from highly talented candidates; and financial savings from cost of absenteeism and presenteeism, cost to cover absent employees, cost of turnover, healthcare premium, and financial returns from more innovative items produced by healthy employees (**Figure 1**). This section explains the role of these KPIs and focuses on principles and basis of strategies to achieve each KPI in workplace health.

3.1.1 Physical fitness

The physical fitness (PF) dimension focuses on the workplace environmental features that contribute to promoting and maintaining physical fitness by creating physical movements and providing spaces for physical activities in the workplace.

This is parallel to the principles of active design that recognizes the built environment as a critical venue for promoting physical activities through the design and layout of the physical environment [16]. Established to develop built environments to increase physical activities to stop the rising level of obesity in the UK, the major agendas of active design involve active travel (or transport) that promotes the use of public transport; active buildings and interiors that integrate physical activity-promoting designs of buildings and indoor spaces; and active outdoor and public spaces that increase physical activities [17]. As sedentary postures and inactivity are prevalent in contemporary workplaces, various design and spatial resolutions can be integrated to promote physical movement and activity.

Interior active design offers various health-promoting opportunities for workplace health via interior spatial planning and design elements, behavioral change encouragement, and available physical exercise spaces. Interior layout and circulation planning can stimulate physical movement by providing choice of spaces for people to work from other types of workspaces than only individual workstations. These choices allow them to break their sedentary workstyle and more frequently move around. It can also incorporate design strategies for the mystery patterns to evoke a sense of mystery to compel employees to explore and move around in the spaces (**Figure 2**). As one of the predictors of the environmental preferences, a sense of mystery triggers a strong pleasure response within the brain comparable to the response in an anticipatory situation [18], which leads to curiosity and interest to further explore spaces that evoke a sense of mystery. A good strategy to creating mysterious environments is obscuring spatial boundaries and a part of the focal object [19, 20].

Interior stairs use is a well-known method for interior active design. The use of interior stairs can facilitate more frequent physical movement in the workplace via three elements: stair design, visibility, and accessibility. People are more likely to be motivated to use stairs instead of elevators when stairs are designed as a focal point with pleasant design elements; clearly visible from the main entrance and even located before elevators; and easily accessible from the major workspaces and

Figure 2.
The American Society of Interior Designers (ASID) Headquarters, Washington, DC, USA. LEED platinum-certified and WELL-platinum-certified. Photo credit: Eric Laignel.

during the regular business hours to all employees. In addition, the use of signage/message board can be an effective means to encourage employees to take stairs instead of elevators by especially posting it next to elevators. A post motivating people to use stairs can significantly increase the use of stairs, four times higher likelihood of using stairs when stairs are visible [21].

Using interior design elements is a great way to motivate employees to be willing to walk more often to other parts of workspaces by creating a pleasant ambience using visual and auditory elements around stairs: art and installations; natural elements such as green plants, plant wall, water feature, or daylight; and pleasant music or nature sounds. Lastly, providing the onsite fitness space and equipment would be a vital way to encourage people to be engaged in physical exercise and activity in the workplace. Onsite physical exercise activities can be further enhanced if a combination of exercise equipment are available for both cardiorespiratory and muscle-strengthening exercises in the space. The WHO recommends adults between 18 and 64 years old be engaged in a minimum of 150 min of moderate aerobic activity (or a minimum of 75 min of vigorous aerobic activity or a combination of both) and a minimum of two times of muscle-strengthening activities a week [22]. When providing a fitness/exercise space and equipment, it is also necessary to provide shower and locker facilities for people to be able to clean themselves after workouts.

Active workstations such as treadmill desks or portable desk pedals, a combination of a desk and a treadmill or a bike pedal to facilitate physical movement while working, can also be integrated to encourage physical movement. However, it is important to note that this is not a means for fitness exercise or weight control. While those who use active workstations are overall more engaged in physical movement than those who do not use them as they tend to break more often sedentary workstyle habits, the amount of exercise is neither enough to meet the recommendations of the WHO as a fitness activity nor to lose weight as a weight control mechanism [23].

3.1.2 Physical comfort

The physical comfort dimension addresses comfort issues in major human senses and body systems. This includes comfort issues in four major senses: *visual, auditory, thermal*, and *olfactory* as well as in the overall body system: *Ergonomics*. Human comfort issues have always been the focus of green buildings. Especially, lighting (visual sense), acoustics (auditory sense), and temperature and humidity (thermal sense) have led major discussions in providing human comfort in green buildings. Smell (olfactory sense) is recently recognized as a human comfort issue in the workplace due to not only unpleasant odors but also recent awareness on toxic chemicals embedded in everyday products used in the workplace such as air fresheners or scents.

Visual comfort strategies for workplace health mainly address lighting for visual acuity, glare control, and quality of lighting. To provide proper lighting for visual acuity, it is necessary to provide sufficient lighting level, 300 lux for general areas and 300–500 lux for computer workstations [24]; control options for light level such as task lights for individual desks and multi-zone lighting control systems with a minimum of several different lighting levels; and availability of light switches and controls within the same space as the lights to be control. Glare control for interior ambient lighting can be addressed through types of lighting fixtures by employing light diffusers to spread the angle of beams wider and indirect up-light fixtures to point lights upward; lighting planning that orient lighting fixtures perpendicular to individual workstations/computer monitors to prevent potential hot spots and glare in monitors; and a coordination with interior finishes as non-reflective matte, light to medium colored interior wall finishes helps to address the overall discomfort with artificial lighting for computer users [25].

Solar glare control strategies have been frequently emphasized in green buildings. Solar glare has become problematic in green buildings as an increased amount of vision glass, glass lower than 7 feet, has been used in office buildings to introduce a higher amount of daylight into the workspaces, coupled with a design intent relevant to the architectural esthetics of glass towers which has driven higher window-to-wall ratios (WWR) in office buildings. Several major strategies to control solar glare include incorporating window shading devices such as interior blinds or curtains, exterior shading systems, variable opacity glazing, interior light shelves, and reflective film of micro-mirrors on windows; meeting WWRs between 30 and 40% [26]; and limiting the amount of direct sunlight that can potentially cause glare discomfort by calculating/simulating the Annual Sun Exposure (ASE). ASE measures the floor area that receives minimum 1000 lux from the sun for at least 250 occupied hours a year, and the recommended ASE is no more than 10% of the total floor area for visual comfort [27].

The strategies to achieve the quality of lighting involve providing proper color rendering of light bulbs through color rendering index (CRI) for the realism of colors in objects; light reflecting property of interior finish materials with light reflectance value (LRV); and brightness contrast level between monitors and immediate surroundings including desk surfaces. The European Standard (EN 12464-1) recommends a minimum CRI of 80 for R_1-R_8 colors for architectural lighting in offices. While only R_1-R_8 colors are considered for CRI ratings, the importance of considering R_9 in selecting architectural lighting has been emphasized as many LED lights in the market have shown poor scores in R_9. With the technological advancement in LED lighting, it is advised to consider various color rendering-related factors simultaneously including fidelity, gamut and correlated color temperature of a light source. LRV recommendations for interior finishes in the office environment vary between the standards and guidelines; more narrowly ranged in the recent green building and high-performance building standards and guidelines for the purpose of energy conservation with a range of 80–90% for floors, 50–70% for walls, 50% for cubicle partitions, and 20–25 for floors [28, 29], but more broadly defined as 70–90% for floors, 50–80% for walls, and 20–40% for floors for visual comfort, specifically, in the office environment [30]. However, LRV is not only related to the amount of light available in the space but also visual ergonomics for visually impaired. It has been noted an association between a lack of contrast between interior finish materials and increased visual fatigue as well as loss of concentration [31]. Visual disability codes and guidelines prescribe the contrast ratios of architectural surfaces and details as follows: a LRV difference of 30 points between two contrasting areas in the British Standard: BS 8300:2009 + A1:2010 and a contrast ratio of minimum 70% between the LRV of the lighter area and the LRV of the darker area in the Americans with Disabilities Act (ADA). Lastly, brightness contrast ratio not exceeding 1:3 should be provided between the monitors and desk surfaces and between the immediate surroundings and the background areas in the office environment to prevent visual fatigue and eye strains [32].

Auditory comfort has been a major discussion in the contemporary workplace. Open-plan offices have become a prevalent office type in the contemporary workplace due to efficient real estate management and encouragement of interactions and collaborations among employees. The importance of open environments has been emphasized to facilitate interactions and collaborations in the innovation economy. The innovation economy is a relatively recent economic theory that emphasizes economic growth through entrepreneurship, technological interventions, and innovations [33]. As seen in successful innovative companies, an open environment that facilitates interactions for fruitful collaborations and innovative outcomes has become the most frequently adopted office type in the innovation economy. However, the downside of an open environment is excessive noise that causes annoyance, stress, and physical fatigue, not to mention loss of concentration and productivity. As noise means unwanted sounds by its definition, when and what is noisy is subjective by individuals based on their workstyles, required tasks at the time: concentration vs. casual mode, and personality attributes. Personality attributes play a significant role in the noise sensitivity issue as people with neuropsychiatric personality tend to be more annoyed by noises and negative people are more significantly influenced by noise annoyance [34, 35]. People who are less extroverted and conscientious might also be more associated with noise sensitivity [36]. *Auditory comfort* in the workplace can be achieved by four major strategies: space planning principles; technical measures; sound masking systems; and sound mitigating construction methods. While many technical strategies and measures are suggested and relied on to mitigate noise issues and provide speech privacy, providing *auditory comfort* in open-plan offices is not an easy task by all means. Thus, these strategies must address a comprehensive approach through space planning and design, technical measures, and workplace policy for successful *auditory comfort*.

Space planning principles provides a basis of noise control in the workplace and include grouping similar types of spaces; placing buffer spaces to separate noisy spaces; avoiding room shapes causing sound to reflect or focus in specific spots; staggering doorways to avoid a straight path for noise; and placing quiet spaces away from noise sources such as major traffic roads and copy rooms (**Figure 3**). Various technical measures can be employed for internal noise control via three typical principles of soundproofing: sound absorption, sound blocking, and sound masking. These technical measures include sound absorption coefficient, noise criteria (NC), reverberation time (RT), sound transmission class (STC), and sound masking systems in open-plan offices. Sound absorption coefficient is a measure that is based on the sound absorption of materials and can assess sound absorption of either individual materials: noise reduction coefficient (NRC) or rooms: design coefficient (DC). As a rule of thumb, a NRC of 0.7–0.8 in ceilings and walls/partitions is considered as a good solution for private offices and meeting/conference rooms, and a NRC of 0.9 or higher for spaces requiring a higher level of speech privacy and open-plan offices [28, 37]. However, the total coverage of the materials applied in an area would determine the overall performance of sound absorption in the space. In addition, the NRC of ceiling materials will be more important than the NRC of wall materials in open-plan offices. It should be noted that NRC is to be replaced by a similar rating system called sound absorption average (SAA). DC is also related to sound absorption of materials but rather assesses the overall sound absorption of a room by considering both the area and NRC ratings of materials in the area.

NC is a sound pressure-based measure and typically used to measure the background noise such as the heating, ventilating and air conditioning (HVAC) systems as this type of noise can be annoying and disruptive in the workplace. A NC of 45-40 is considered appropriate for open-plan offices, multi-occupant spaces and common spaces; a NC of 35-30 for conference and meeting rooms with normal speech privacy, a NC of 25 for conference and meeting rooms with high speech privacy, and a NC of 20-25 for teleconference rooms [37, 38]. According to Noise Rating (NR), a NC equivalent in Europe, NR 35 is recommended for open-plan offices and 30 for general offices. RT in a room is measured by the amount of time it takes for sound to decay by 60 decibels (T60). A RT of 0.5–0.6

Figure 3.
Interface Headquarters in Atlanta, Georgia, USA. A combination of acoustic solutions used: separate acoustic zoning, quieter background noise from chilled beam system, mechanical sound tuned to sound masking pitch, and interior finishes and furnishings with acoustic properties. LEED platinum and WELL silver certifications pending. Photo courtesy: Nick Merrick, Hall + Merrick photographers.

is considered for private offices, meeting/conference rooms, and teleconference rooms, and a RT of 0.8 for open-plan offices [37, 39]. The overall performance of RT in a room will be determined by sound absorption of materials and the volume of the space as it gets longer with less sound absorption and a larger volume.

STC, as a laboratory test - based rating for sound blocking and isolation, measures the degree of sound attenuation of a wall partition in adjoining areas, commonly ranging between 30 and 75 [40]. STC 40 is considered a threshold for a sense of privacy and higher numbers are required to provide speech privacy and noise isolation. A minimum of STC of 45 is required between standard offices to block loud speech and 50 between meeting rooms, conference rooms, and hallway to isolate louder sounds [28, 39]. Sound Reduction Index (Rw), instead of STC, is used in Europe to measure sound insulation levels by calculating the reduction in the intensity of sound through a partition. Rw 45dB means when loud speech is audible but unintelligible through a wall and, RW 50dB and over when loud speech is difficult to be audible. A field test version of Rw, Sound Level Difference (Dw) with NR together is often used to determine Acoustic Privacy in Europe. A privacy rating of 75 is considered as a good acoustic privacy level and, 85 a high level of acoustic privacy where loud speech is barely audible and intelligible. Sound masking is a vital acoustic system for sound privacy in open-plan offices by adding background noise. An electronic sound masking system typically used in open-plan offices generates ambient background noise that matches the frequency of human speech. This makes conversations in open environments unintelligible to nearby coworkers, providing sound privacy. Various masking sounds such as white noise and pink noise have been explored to reduce the intelligibility level of speech from adjacent co-workers in open-plan offices. While high acoustic variation such as music can affect emotion positively, it tends to distract people's concentration and disrupt with short-term working memories and cognitively demanding functions. Among nature soundscapes, water sounds with limited acoustic variation have shown more effectiveness than high acoustic variation such as bird songs. Sound mitigating construction methods help decrease sound travel and increase sound isolation via the details during the construction, including constructing interior walls/partitions with staggering gypsum board seams; completely sealing interior walls at the top and bottom; using a door with a non-hollow core, gaskets all around door perimeters, and door sweeps for meeting rooms, conference rooms, and private offices; using sound isolation hardware, such as resilient channel clips or floor isolation hardware.

Thermal comfort is the source of one of the most frequent complaints in the office environments. In periodic facility satisfaction surveys, being hot and being cold have consistently been top two complaints about office environments [41]. *Thermal comfort* is one of the most difficult challenges in meeting expectations of occupants as it is determined by (1) environmental factors: air temperature, air velocity, radiant temperature, and relative humidity; (2) personal factors: clothing and metabolic rate, and (3) work-related factors including work demand. In addition, other biological and physiological factors also affect thermal sensation among various people. Women tend to prefer higher temperatures, preferably 2.5°C higher, than men [42, 43] and are more sensitive to temperature than relative humidity while men are the opposite [44]. Similar to noise sensitivity, some people have thermal sensitivity, showing lower tolerance to undesirable thermal conditions. Due to individual and biological differences in thermal sensation, thermal comfort in the workplace needs to be sought in spatial planning, work settings, and policy support beyond technical and mechanical approaches.

Thermal comfort in the workplace can be addressed via fundamental thermal comfort measures and advanced temperature control through ventilation systems. Fundamental thermal comfort measures focus on thermal comfort issues in an integrative way by considering environmental factors, flexible work settings, and policy together: providing individual thermal comfort devices such as fans and protective clothing or blankets; allowing a flexible workstyle for employees to be able to move to another space with a desirable temperature or shift to another type of task requiring a different level of metabolic heat; providing spaces with various temperatures for employees to choose from; providing cabinets, lockers, or cloak room for people to store extra layers of personal clothing or blankets; separating heat-generating equipment and devices from major workspaces; maintaining 30–60% of relative humidity (up to 65% for mechanically ventilated spaces); and policy to conduct an occupant survey to ensure that at least 80% of people at work are satisfied with temperature and humidity as prescribed in ANSI/ASHRAE 55-2013 [45]. Advanced temperature control via ventilation systems can be an expensive investment and, thus, requires consultancy with experts to carefully determine pros and cons of systems, adequacy to address thermal comfort in the particular environments, and the return on investment. These systems include operable windows with outdoor air monitoring system in a mechanically ventilated office; natural ventilation system or passive cooling; radiant heating system; dedicated outdoor air system; and displacement ventilation system such as underfloor air distribution system or low-side wall air distribution system.

Olfactory comfort concerns not only unpleasant odors but also toxic chemicals used to create pleasant odors. Unpleasant odors in the workplace can come from various sources such as food, garbage, bathrooms, and storage. In addition, there is an issue of chemical sensitivity in the workplace for those who are allergic to certain chemicals used in fragrance, scents, air fresheners, and cosmetic lotions. A recent study shows that nearly 20% of people have chemical sensitivity to chemicals used in everyday products and over 90% of people with multiple chemical sensitivity have fragrance sensitivity [46]. This raises a caution to daily chemicals used in the workplace. There is also a general concern with harmful chemicals used in cleaning agents in the workplace. To prevent strong odors from the spaces where the sources exist, the following strategies help to alleviate the problems and enhance *olfactory comfort*: providing self-closing doors; deck-to-deck partitions; separate exhaust; negative pressurization; a buffer space such as an interstitial room, vestibule or hallway separating spaces generating strong odors from major workspaces. In addition, having a workplace policy regarding fragrance sensitivity and the use of other daily products with toxic chemicals such as air fresheners is a good way to enhance *olfactory comfort*.

Lastly *furniture ergonomics* provides a support for the overall physical bodily comfort. Due to the efficiency-oriented work settings and longer working hours in the contemporary workplace, a sedentary workstyle has caused adverse health issues. A significant amount of work-related musculoskeletal disorders (WMSDs) is a well-known issue in the workplace and a critical reason for the need of ergonomic furniture. WMSDs are injuries or disorders in muscles, spinal discs, nerves, tendons, joints, and cartilage caused by the work environment or certain work conditions such as repetitive movement or forced awkward postures [47]. Typical WMSDs include back pain, sprains, strains, tears, and carpal tunnel syndrome. WMSDs are not only causing an annoying pain, that is often chronic, to manage for employees but also a significant amount of costs to employers in the forms of compensation costs, absenteeism, increased healthcare cost, and lost productivity [48]. *Furniture ergonomics* is critical to properly support the human body systems and reduce the risks of WMSDs. Many standards and guidelines related to ergonomics of workplace furniture can be easily found in the literature from the Occupational Safety and Health Administration (OSHA) of the US Department of Labor, the Human Factors and Ergonomics Society (HFES), and the Business and Institutional Furniture Manufacturers Association (BIFMA).

The fundamental ergonomics for workplace furniture include providing individual workstations with ergonomic sizes and clearances; ergonomic conditions of individual workstations in open-plan offices; and ergonomic chairs for individual workstations. Ergonomic sizes and clearances of individual workstations include: a minimum desk surface of 20 inches (50 cm) between a seated person and a monitor and a minimum of 30 inches (76 cm) to place both a keyboard and a monitor; height-adjustable desk between 20 inches (50 cm) and 28 inches (72 cm) for seated tasks; height-adjustable desk to accommodate standing tasks; sufficient under-desk clearance, minimum width of 20 inches (50 cm) and a minimum depth of 15 inches (38 cm) for knees and 24 inches (60 cm) for feet. Ergonomic conditions of individual workstations should address: perpendicular placement of desks to window panes to reduce glare and brightness contrast issues; matt finish on desk surfaces to reduce glare; 24–27 inches wide desk surface edge to accommodate armrests of chairs; and round front edge of desk to avoid contact stress of wrists. Ergonomic chairs for individual workstations should have such attributes as height- and depth-adjustable seat; seat with a minimum width of 18 inches (45 cm) and the length between 15 inches (38 cm) to 17 inches (43 cm); rounded, waterfall edged front of the seat; height-adjustable armrests; armrests with tilting capability (inward and outward from the user); backrest with reclining capability, a minimum of 15 degrees; backrest with a minimum height of 15 inches (38 cm) and a minimum width of 12 inches (30 cm) with a lumbar support; and chairs in different sizes/fit available for people with a particularly small height (5′2″ and lower) or tall height (6′3″ and above).

3.1.3 Physical nourishment

The science clearly tells us how physical activity and diet issues go hand-in-hand together in human health. According to the WHO, poor diet, especially, consuming a significant amount of foods and beverages high in calories but low in nutrients, is the main culprit of overweight and obesity along with physical inactivity [49]. A medical condition called metabolic syndrome or insulin resistance syndrome, a combination of metabolic disorders showing the symptoms of obesity, high blood pressure, and diabetes is a particular concern of public health related to chronic diseases. Dietary factors significantly contribute to both the development and control of metabolic syndrome. Metabolic syndrome as well as cardiovascular disease (CVD) and certain cancers can be improved and prevented by certain dietary factors, including consuming healthy fats, consuming fruits and vegetables, eating whole grains, and limiting sugar intake and overall excessive calories [50]. In addition to dietary factors, professionals in the workplace design, planning, and management have also started developing strategies through environmental attributes, choice architecture, and workplace policy to support healthier diet and behaviors in the workplace. In this capacity, the built environment professionals are expanding their traditional roles to facilitate and orchestrate the overall workplace health, working with other professionals in nutrition, diet, genetics and neuroscience, as well as human resources to tackle such a complex problem through multi-disciplinary interventions. This section focuses on providing *healthy food amenities* as environmental strategies and encouraging *health-conscious eating habits and behaviors* through the method of choice architecture.

The strategies to support healthy eating via *healthy food amenities* include providing onsite food-related amenities for employees; and enhancing onsite eating space features. Onsite food-related amenities include cold food storage, cabinets and storages available for personal food items that employees bring, food heating-up appliances such as microwave, and dish washing supplies and sink. Enhancing onsite eating space features encourages employees to use the space to prevent eating alone or at individual desks and rather socialize with colleagues during lunch time. Eating in a social setting has several benefits. It slows eating speed down, preventing people from eating too fast and consequently overeating. It also makes people more mindful of the type, amount, and contents of food they consume, compared to eating alone or eating while working. And there is a benefit of increasing social well-being while conversing and socializing with others. Desirable features of eating spaces include various types of seating arrangement available; daylight; art and design elements available such as paintings, plants, water features, or music; and being sufficiently large enough to accommodate at least 25% of employees at once.

The main strategies to encourage *health-conscious eating habits and behaviors* involve the concept of choice architecture. Choice architecture is a practice of presenting choices to consumers at points-of-decision with strategies to influence their decision toward desirable options in behavioral economics. This influence on people's decision is called nudging. Choice architecture has been practiced in commercial food industry for a while. The same principles and strategies are now suggested to influence people to make healthier food choices. These strategies can be divided into two groups: food-related and non-food-related. Food-related choice architecture strategies include proximity of healthy foods—placing healthier food choices close to people's point-of-decision or purchase point; portion control—smaller portioned food options available; promotion of healthier choices—use of more appealing packing, appearance or display of healthier foods; and healthy default—providing healthy default side dishes with bundled meals. Non-food-related choice architecture strategies are ambience—using environmental design elements, such as lighting, colors, or spatial features, to highlight healthier food choices; pricing—providing reduced prices for smaller portioned options or healthier options; prompting—providing nutrient information and use of messages or signage at points-of-decision or purchase point; and convenience—providing an expedited checkout options for foods marked as healthy food options.

3.1.4 Cognitive well-being

The cognitive well-being dimension emphasizes environmental supports to enhance cognitive mental processes of attention, memory, problem-solving, reasoning, and thinking for occupational functions. Cognitive domain mainly focuses on the mental capacity related to thinking process, acquiring knowledge and associated behaviors, while emotional domain focuses on the feelings that affect moods, attitudes, and consequently behaviors. However, human mental domain is a complex territory that is intertwined and influential among the components within themselves. It is often seen that a positive emotional mood is associated with enhanced cognitive performance. Thus, cognitive well-being in this section mainly concerns on how to enhance mental capacity related to job functions through the environment, and the emotional well-being section, on workplace environmental attributes to create positive moods and attitudes in its own dimension.

A cognitive performance-conducive environment in the workplace is an age-old discussion since job performance in the contemporary knowledge workplace is mainly related to employees' cognitive functions. Many studies in organizational behavior and management have explored how to create work environments conducive to employees' creative and innovative ideas and problem-solving under the innovation economy. As entrepreneurship, technological interventions, and innovations have become important for economic growth under new financial challenges and downturns, specific workplace strategies have been sought to enhance employees' cognitive performance in the knowledge workplace. Cognitive performance and well-being has also been discussed in cognitive and organizational ergonomics. As a study of human-system interactions at work, when applied to the environment, cognitive ergonomics focuses on environmental supports for mental process of information and memory, while organizational ergonomics, on organizational structures and policies [51]. Cognitive well-being strategies comprise three major themes to reduce cognitive overload and increase support systems: providing appropriate *types of spaces*; *flexibility and flow* in the major workspaces; and *acoustic privacy*. In addition, crowding and density, proximity, and visuospatial information process affect cognitive performance and well-being at work. Thus, environmental supports are necessary to provide appropriate space planning and sensory balanced environments to reduce cognitive overload, stress, and errors, and assist with better attention and decision-making.

According to the literature in organizational management, choices and autonomy in the workplace contribute to organizational creativity and innovation by building trust, transparency, and even engagement. Allowing employees to be able to find appropriate *types of spaces* for the tasks that they need to accomplish is critical to support their job functions. There are four types of work modes in the knowledge workplace including socialization, externalization, combination, and internalization [52]. Typical spaces necessary to support these four types of knowledge work modes are spaces for focus, socialization, and collaboration. Collaboration spaces can accommodate various needs and modes of collaboration such as formal vs. informal and scheduled vs. impromptu. These collaborative spaces can also be either technology-enhanced in various degrees or technology-free depending on the purpose of the tasks and outcomes. When providing these spaces, it is important to offer various degrees of settings for sensory balance. For instance, the workplace can identify best settings for spaces with higher privacy vs. spaces for interaction, quiet spaces vs. noisy spaces, and various lighting levels and thermally variant spaces to accommodate physical comfort to enable their best cognitive performance. Along with these spaces, various furniture settings can also be provided: formal setting vs. informal/casual settings, and high ergonomic settings for focused mode and longer stay vs. low ergonomic settings for casual and shorter stay.

Respite or recharge spaces are a fourth type of spaces necessary for employees to mentally cope with stress, work demand, and physical fatigue. Contemporary workers are frequently cognitively overloaded, receiving more information every second than their conscious brain can digest. This causes a constant mental fatigue and blocks that adversely affect their performance. There are multiple studies that highlight the benefits and the power of breaks and respite on physical and mental restoration. When taking micro-breaks, especially, engaging in relaxation and socialization activities throughout a day, employees tend to be less stressed by work demands and better recover from negative affect from work-related stress [53]. They also tend to have a higher level of focus, a greater creative thinking capability, and a substantially increased sense of health and well-being [54]. Respite or recharge spaces, intending to provide a short mental break and cognitive restoration, consist of play spaces, solitude spaces, social spaces, and designated outdoor respite spaces (**Figure 4**). Play spaces are provided for physical activities and games with supportive equipment. Solitude spaces are for personal lounging, reflection, contemplation, meditation, or religious practice. Social spaces are for social interactions and gatherings. Designated outdoor respite spaces include outdoor lounge areas, outdoor garden, balconies, or rooftop hangout places. If there are not designated outdoor respite spaces associated with the workplace, having access to public respite spaces is another great way to offer employees outdoor respite spaces.

Flexibility and flow of major workspaces are necessities to supporting cognitive functions by providing flexible spatial capacity and flow to aid various tasks and numbers in team/group as well as to capture the flow of ideas. The success of a knowledge workplace depends on successful organizational knowledge management. And knowledge management ability relies on how an organization can capture, share, manage and deliver information in real time to make faster but accurate

Figure 4.
Interface Headquarters in Atlanta, Georgia, USA. A respite, social space with biophilic/evolutionary psychological design features, highlighting high IAQ and non-toxic materials. LEED platinum and WELL silver certifications pending. Photo courtesy: Nick Merrick, Hall + Merrick photographers.

decision-making [55]. *Flexibility and flow* of major workspaces can be achieved by four strategies: providing expandability, versatility, and convertibility of major workspaces, and implementing non-hierarchical spatial planning. Expandability addresses a capacity to accommodate growth with expansion such as a space with movable partitions to open up the space to accommodate bigger groups. Versatility refers to a capacity to accommodate several different activities as multifunctional spaces such as a space with various settings of furniture to accommodate different activities in the same space. Convertibility means a capacity to change in function through the conversion of spaces such as a space that can easily be modified to change its purpose to serve another function. Non-hierarchical spatial planning is spatial planning that blurs the traditional power structure at work through different space allocation or layout from the traditional planning such as assigning spaces by the best use of the spaces or types of furniture by the functions of the person instead of the hierarchy of the organizational structure. Non-hierarchical organizational structure tends to promote an organizational culture that encourages individual views and participation, informality, and freedom, which is crucial to innovative performance and outcomes [56].

Acoustic privacy is one of the major problems in open-plan offices. *Acoustic privacy* can be achieved through spatial options and behavioral change: providing separate, designated focus/concentration spaces away from open workspaces; providing a phone booth type of small private rooms for confidential or private conversations or calls; conducting an early assessment, identification, and space planning to establish quiet zones separately from noisy/interaction zones; and workplace policy for office etiquettes. It should be noted that the major sources of noises in the workplace are people-related issues. People's conversations with others nearby or via phone and telephone left ringing being most disruptive, significantly worse than noises from office equipment and outside the building [57]. Thus, workplace etiquettes promoting courteous behaviors for others who might be sensitive to noises can be an impactful way of mitigating noise issues at the root of noise sources through workplace policies and achieve *acoustic privacy*.

3.1.5 Social well-being

The social well-being dimension focuses on enhancing *social connectivity* and support social systems in the workplace. While many conversation in the built environment revolves around physical and mental well-being, research suggests that social well-being has huge impact on both psychological and physical health as well as longevity [58, 59]. The intangible impact of social well-being can be even more significant, affecting longevity more than smoking, drinking, and obesity [60]. Thus, it is important to address how to promote *social connectivity* via workplace design and planning. Spatial layouts significantly affect how people interact and have been used to predict social interaction patterns and behaviors in the built environment.

Providing *social connectivity* between people in the workplace through space planning and design features can enhance not only social bond and inter-personal relationship between colleagues but also interactions and communications necessary for job functions. When communications increase in the workplace, people tend to also talk about task-related conversations, which then promote feedback, conflict reduction, and increased motivation [61]. Socialization activities at work have shown to reduce work demand-related stress and negative affect at the end of workday [53]; increase cohesion; and decrease employee turnover [62].

Enhancing *social connectivity* through spatial layout and planning can be done in four ways: providing social spaces; increasing visual connectivity; establishing core interaction networks; and implementing cross-pollination space planning practice. Social spaces include impromptu meeting spaces at corners, along the hallways or major paths of traffic to increase serendipitous interactions and meetings; central or local open lounge areas; food/beverage spaces such as café, snack bar, or pantry; and breakrooms. Visual connectivity can be achieved via either horizontal or vertical visual connectivity. Horizontal connectivity can be enhanced through the use of lower partitions than 42 inches (1 m) high or transparent materials for partitions and enclosed spaces. Vertical connectivity can be achieved by creating a vertical openness using inter-floor stairs or atrium. Establishing core interaction networks is a vital way to increase social connections. An organization is defined as connected points of people that create the pattern of communication or information exchange among them, acting as a network in an organization, and social networks play a critical role in generation innovative outcomes [63]. Core interaction networks can be established by placing individuals who are the core interaction networks within a department/team along the core circulation paths for easy access and increased exposure for others. Furthermore, cross-pollination space planning practices can promote social interactions by connecting core circulation paths to the informal or impromptu meeting and social spaces where colleagues in different departments/teams can frequently meet and gather informally.

3.1.6 Emotional well-being

The emotional well-being dimension mainly concerns workplace environmental features providing individual emotional state of happiness and satisfaction. Emotional well-being is a dimension above the first two basic human needs: biological and physiological need and safety need in Maslow's hierarchy of needs that explains behavioral motivation [64]. Emotional well-being is a critical dimension to the overall quality of life, and can be addressed via three major topics related to environmental features in the workplace: *biophilic design*, *art and design elements for pleasure*, and *personalization and control*.

Biophilic design refers to environmental design strategies and features to enhance human's innate bonding with nature. It has been long recognized human's strong instinct responsive to natural forms and patterns, and their desire to integrate those into their habitats and lifestyle. Restoring this bond, once broken due to industrialization and modernization, is now emphasized as a critical component to human mental well-being. Nature has been highlighted for its therapeutic benefits on improved moods, reduced stress, and restoring human adaptive resources [65–67].

Biophilic design strategies consists of three major topics: providing sensory stimulation by nature and integrating natural elements; views and circadian lighting support; and accommodating evolutionary psychological design.

Sensory stimulation by nature and integrating natural elements are related to visual, auditory, and olfactory sensory experiences with natural elements as well as their components expressed in design elements. There is a plethora of studies that provide evidence of multiple benefits from the power of seeing or contacting with nature on human mental capability coping with physical healing, stress and fatigue reduction, and faster cognitive restoration. In addition, using vegetation or plants in interior spaces can be an effective strategy to reduce noise, provide privacy, and decrease negative feelings of crowding, which also contribute to health and well-being in the workplace. Recently, organizations have started paying attention to the positive relationship between natural elements and employee's enhanced cognitive function and performance, and, thus, potential organizational benefits from increased productivity as well as engagement. Providing sensory stimulation by nature and integrating natural elements can be achieved by presence of green vegetation; water elements; 2D or 3D artwork depicting natural elements or scenes; sounds of nature; smells of nature; shapes and forms of nature in workspaces or materials; patterns of nature applied to the spatial design; materials of nature; colors of nature.

Circadian lighting is a recent focus among lighting professionals to provide human-centric lighting by supporting human innate body clocks, circadian rhythms, that synchronize human physiological functions with the approximate 24-h day-night solar cycle. As a type of neuron in the retina of eyes, the intrinsically photosensitive retinal ganglion cells (ipRGCs), sends information on the time of day to an area in the brain called the suprachiasmatic nucleus which synchronize clocks in peripheral tissues and organs in human body, the role of light has become more than providing visual functions. Circadian lighting is known to regulate various hormones such as growth hormone, melatonin, and cortisol that affect sleep, mood, stress, and eating and diet habits. Imbalance in circadian rhythms can cause adverse health consequences such as sleep disorder, obesity, diabetes, cardio-metabolic diseases, and even memory and productivity [68–70].

Views to outdoors and circadian lighting support can be incorporated through non-obstructive views and no tints on the windows; minimum two lines of sights to vision glass in the primary workspaces; views of natural elements outside the windows such as plants, skies, etc.; non-obstructive views within the distance; locating at least half of the primary workspace (or regularly occupied spaces) within 40 feet (12 m) of windows or atriums. Furthermore, using a technical metric, either circadian stimulus (CS) or melanopic lux (ML), to assess the amount of circadian light can assist the enhancement of circadian impactful lighting (**Figure 5**). However, when it comes to design solutions for circadian lighting, a cautious approach is necessary. While there is a sufficient amount of research indicating the impact of light on our body clock, it is still too early to be conclusive on what solutions are truly desirable and effective. In addition, daylight is a complex issue as it can create other undesirable problems: glare, overheat, and higher cooling loads. It is worth noting an approach suggested by the US General Services and Administration (GSA) called daylight ecosystems that require a comprehensive approach to daylight solutions by integrating daylight design for daylight penetration strategies, interior design for choices of finish materials and desk orientations, and human factors to support user behaviors [71].

Evolutionary psychology is an approach to the study of human behaviors that explains how the brain works and behaviors form based on biological reasons. Prospect-refuge theory, similar to mystery theory discussed in the active design section, explains many aspects of human visual and environmental preferences in nature and their psychological comfort, security, and safety related to these preferences. According to the theory, people feel secure, safe, and comfortable in a space with an open view: prospect for easy surveillance to detect hazards, and a space protected from behind and above: refuge to safeguard themselves from unexpected hazards. Prospect-refuge theory has many implications especially in open-plan offices. These implications include visual access to activity hubs or similar central spaces from individual workspaces with lower partitions and transparent materials; dropped ceiling and/or elevated floors in areas with high ceilings; quiet spaces covered by three sides such as reading nooks, booths, or alcoves; operable or adjustable translucent screens or semi-opaque partitions; clusters of workstations; freestanding planters or vegetated partitions in circulation paths between clusters of workstations; and retreat spaces looking out to outside views.

Art and design elements for pleasure addresses human innate desire for esthetic pleasure. Esthetics used to be a domain of psychology but neuroscientists are currently also exploring the answers for why humans develop esthetic adjustment and appraisal by examining the human brain. The approaches to *art and design elements*

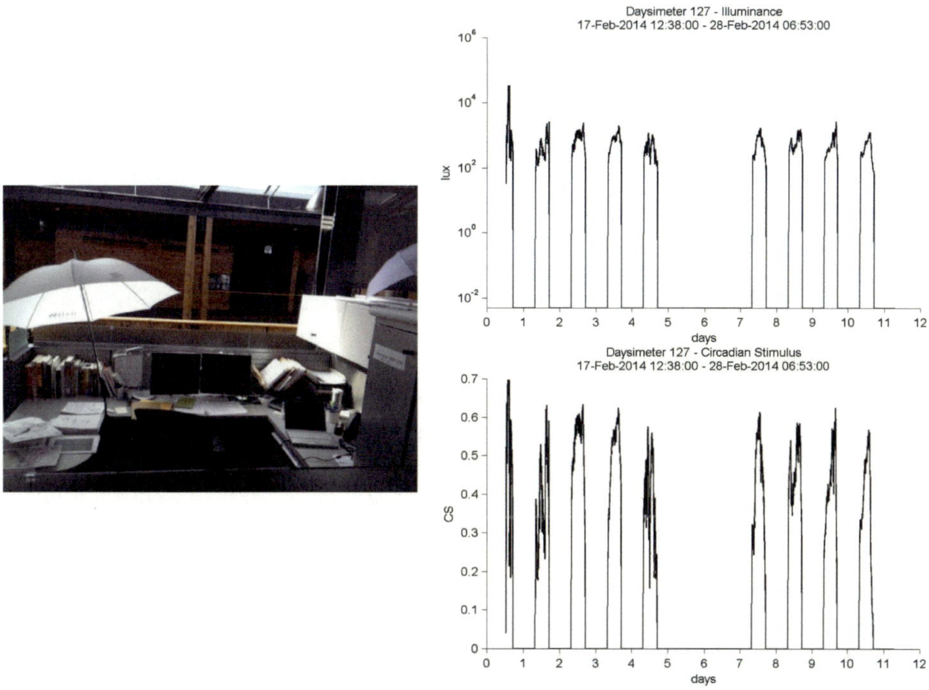

Figure 5.
Circadian stimulus measured at a desk with an umbrella improvised by employee to block too much sunlight and glare from the skylight, the US GSA Federal Center South building. Photo courtesy: the US General Services Administration.

for pleasure are composed of three topics: incorporating art and design elements; sensory engagement in common or open spaces; and relaxing settings.

Art is known to have not only emotional impact but also cognitive and psychological stimulation. In addition, the literature suggests its healing power shown through positive physiological improvements related healthy symptoms [72]. Art and design elements deal with conventional artwork and installation as well as unconventional design components in the workplace. According to the organizational management literature, ease and laughter in the organizational culture in the workplace is one of the critical elements contributing to less stress, cognitive release, and social cohesion. These concepts can be expressed by fun, whimsical, unconventional design features that we have witnessed in successful tech companies and creative industries as part of their effort to create an ambience of ease [73]. Art and design elements for pleasure include artwork, designwork or installation in major common spaces and regularly occupied spaces; interior components with unconventional functions and shapes; unconventional use of finish or furnishing materials; and fun and unique decor such as whimsical signage, ornaments, or similar items.

The sensory engagement in common or open spaces involves three common sensory systems: visual, auditory, and olfactory as well as another sensory system that is typically expressed in interior design elements: tactile. Sensory engagement comprises visual elements utilizing colors and patterns; auditory elements, music or sounds of nature; olfactory elements, smell or scent such as smell of coffee or herbs; and tactile elements, various levels of soft and hard textiles and textures. These sensory stimulations tend to influence emotional states and also raise mental awareness and cognitive engagement. There are, in general, more design strategies to create visual stimuli utilizing colors and patterns than other types of sensory stimuli. This is due to the human neurophysiological condition that 30–40% of human cerebral cortex is allotted to vision, making vision the most dominant sense in the human body, while the cerebral cortex for auditory sense is only 3%.

Relaxing settings create an ambience of positive state of mind and ease. Relaxing settings are related to expanded experiences in the workplace that are associated with the first place (home) and the third place (café) beyond the second place (workplace). The attributes of the first place bring psychological comfort to the workplace, and the attributes of the third place, a sense of community and social activities as an anchor in the community. Home-like settings, especially, enhance such organizational cultures as freedom and autonomy which contribute to flattened hierarchy and nonconformity that are critical to employees' mental well-being. In addition, certain successful start-up companies have seen a significant role of such a home-like environment as a work environment in increased creative and innovative outcomes. This is because such an environment enhances organizational cultures vital to business success such as less hierarchical structure, more social connections between colleagues, and shared visions and goals. Sensory engagement in common and open spaces can be enhanced by providing furniture/furnishing types associated with the first place or the third place; and spaces emulating residential setting or the third place.

Personalization and control is an essential component for workplace health in open-plan offices. Numerous studies support the positive impact of personalization of work spaces and furniture arrangement as well as personal control over the environment on job commitment, motivation, engagement, job satisfaction, and performance. Personalization is a critical mechanism for employees to express their emotions, manage stress, and control their workspaces [74]. This tends to be more important to women as women personalize their spaces more than man and more frequently rearrange their workspaces [75]. The strategies for personalization and control in open-plan offices consist of allowing employees to display physical personal items such as artwork, plants, photos of people in individual workspaces; display personal items such as personal photos digitally through computer screensavers; adjust/reconfigure the settings in individual workspaces; and choose the layout of their individual workspaces.

3.1.7 Environmental well-being

The environmental well-being dimension concerns providing non-toxic and clean environment for employee health. There has been a plethora of research conducted in this topic, and many technical standards and guidelines have been established to protect human health in the built environment. While an extensive list can be sought to battle the problems related to toxins, chemicals, and cleanliness in work environments, this section discusses four main sources that are typically found in the workplace: *indoor air quality (IAQ)*; *drinking water quality*; *chemical control*; and *cleanliness and maintenance*.

IAQ is one of the strongest predictors of workplace health as well as performance. Poor *IAQ* has been linked to both physical, psychological, cognitive health symptoms: tiredness, loss of concentration, loss of control, depression, anxiety as well as physical conditions: irritation of skin, respiratory tract, asthma, allergic reactions to pollutants, infectious diseases and even cancers [76]. In addition, poor *IAQ* problems are known as one of the leading causes of distractions and decreased productivity [77]. *IAQ* in the workplace can be controlled, maintained, and enhanced through three themes of major strategies to provide good quality air: removing or preventing chemicals, pollutants, and irritants in the air; providing air filtration system; and employing *IAQ*-enhancing features. To remove or prevent harmful and hazardous toxins and irritants, the following strategies need to be sought. Indoor smoking should be banned. A minimum of 10 feet long entryway system needs to be implemented in building entrances to remove and prevent outside dirt and particulates from entering into the building. Sources of chemicals and pollutants need to be controlled in the janitor closets, bathrooms, copy/print rooms, and storages for products containing harmful and hazardous chemicals by using a combination of self-closing doors, deck-to-deck partitions, exhaust rate of minimum 0.5 cubic feet per minute (cfm) per square foot (0.15 $m^3/min/m^2$) with no air recirculation, and negative pressure at least 5 Pascals on average and 1 Pascal at a minimum when the door of the room is closed.

Volatile organic compounds (VOCs), organic chemicals that easily evaporate at room temperature due to a high vapor pressure, cause irritations in eye, nose and throat, headaches, nausea, and damage to human organs and central nervous system. VOCs must be controlled and limited in various building, interior, and cleaning products. There are various standards and guidelines in relation to VOCs in the built environment that can be complied with when seeking to purchase particular items in the workplace. These include but not limited to the following: Green Seal Standard GS-36 Paints & Coatings, GS-11 Paints, and GC-03 Anti-Corrosive Paints, ASTM D 2369-10 Standard Test Method for Volatile Content of Coatings, ISO 11890 Paints and varnishes—Determination of VOC content, CRI Green Label Plus Program & Green Label Program for carpet, cushions and adhesives, FloorScore for surface flooring materials, adhesives, and underlayments, and ANSI/BIFMA e3-2011 Sustainability Standard Method for furniture.

Air filtration systems are a good means to enhancing *IAQ*. Indoor air pollutants are made of particulate matter typically classified into three groups based on sizes related to their deposit behaviors in human body airways: particulate matter smaller than 10 μm (PM_{10}), 2.5 ($PM_{2.5}$), and 0.1 μm ($PM_{0.1}$). Particles with a diameter greater than 10 μm are less of a concern in IAQ as they tend to be filtered out by human nose and upper airway. Air filter systems in indoors are more concerned with filtering out PM_{10} (coarse particles), $PM_{2.5}$ (fine particles), and $PM_{0.1}$ (ultrafine particles) as they can reach human respiratory system. Air filtration system strategies can be implemented by incorporating in the ventilation system air filters rated with a Minimum Efficiency Reporting Value (MERV) of 13 that filters 90% of PM_{10} and $PM_{2.5}$. The workplace can also employ air purification and disinfection filtration systems to clean indoor air and kill microbes to control pathogens. These systems include carbon filters, ultraviolet germicidal irradiation (UVGI), or photocatalytic oxidation (PCO). Carbon filters are known to filter pollutants and VOCs as well as absorb odors and fumes. UVGI or PCO are recommended for purification and sanitization of air. When employing these filtration systems, it is also important to develop and adopt a policy for a regular filter maintenance program to be able to continuously monitor and sustain the enhanced air quality.

The workplace can further explore other strategies to enhance *IAQ*. Providing a real-time air monitoring display system is a great way for employees to understand the level of IAQ in their spaces by showing the major indoor air pollutants such as carbon dioxide, ozone, and particle count. The workplace can also develop and adopt a green purchasing policy as a procurement policy of products and services to reduce negative impact of *IAQ* on occupant health. Lastly, controlling and monitoring outdoor air delivery and quality can enhance *IAQ*. The workplace can increase outdoor air ventilation rates in all occupied spaces by minimum 30% above the ASHRAE 62.1 requirement and monitor outdoor air contaminant concentrations regulated by National Ambient Air Quality Standards (NAAQS) for the six common air pollutants: carbon monoxide, lead, ground-level ozone, nitrogen dioxide, particulate matter, and sulfur dioxide.

While *drinking water quality* is controlled and managed by the laws and codes for basic human health, there are strategies to further enhance the quality of drinking water in the workplace through water filtration systems. These water filtration systems are employed to clean, sanitize, and screen organic and non-organic pollutants. Sediment filter systems, acting like a net, screen solid particulate transported in water. Reverse osmosis (RO) systems and kinetic degradation fluxion (KDF) filters can remove dissolved hazardous metals such as lead, mercury, and arsenic from water. UVGI and the National Sanitation Foundation filters are sanitization filters to reduce microbes in drinking water.

Hazardous chemicals in building and interior materials need to be controlled. Persistent bio-accumulative and toxic (PBT) chemicals are especially a major concern. PBTs are chemicals that do not easily degrade and cause adverse health impact such as reproductive dysfunction, endocrine disruption, or cancers by accumulating in organisms for an extensive period. There are many green building standards and guidelines that have established what chemicals need to be regulated in building and interior materials for occupant health. The workplace should ensure that there are no human carcinogens such as asbestos, lead, polychlorinated biphenyl (PBTs), mercury, cadmium, perfluorinated compounds (PFCs), halogenated flame retardants, phthalate (plasticizers), isocyanate-based polyurethane, and urea-formaldehyde in building materials; and no urea formaldehyde, mercury, lead, antimony, cadmium, and halogenated flame retardants in furniture and furnishings materials. In addition, the workplace can purchase furniture that is certified by the Scientific Certification Systems (SCS) Indoor Advantage Environmental Certification and interior finishes and furnishings certified by Declare®, Cradle to Cradle®, Cradle to Cradle® Material Health, Health Product Declaration®, Underwriters Laboratories (UL)' Product Lens™, Facts-NSF/ANSI 336, or SMaRT©.

Lastly, *cleanliness and maintenance* should be addressed to sustain the quality of indoor environment that is controlled, monitored, and toxin-reduced. *Cleanliness and maintenance* can be addressed in two ways: cleaning equipment and agents, and cleanliness and maintenance level. The workplace can develop and adopt a protocol for cleaning and maintenance by determining frequency, supplies, equipment, and procedures. Another protocol is also necessary for the cleaning storage to instruct how to handle and store hazardous chemicals such as bleach and ammonia-based cleaning products. Using cleaning agents that are certified by a reliable third party is a good way of practicing how to identify and purchase cleaning agents that do not contain harmful chemicals. These certificates include the US EPA Design for the Environment, UL EcoLogo®, or GreenSeal®. The cleanliness and maintenance level can be addressed by achieving cleanliness and hygiene that is equivalent to Level 2 Ordinary Tidiness as defined in the Association of higher Education Facilities Officers (APPA) custodial guidelines for general office spaces. The Level 2 Ordinary Tidiness indicates when there is no built up in corners or along walls; there is no more than 2 days' worth of dust, dirt, stains or streak; and marks, smudges and finger prints are noticeable only upon close examination. Hygiene-sensitive spaces need additional care. This requires cleaning all countertops and fixtures in the bathrooms and kitchen and shelves and surfaces inside refrigerators with proper hygiene cleaning agents on a regular basis. In addition, a management policy for pest and garbage should be in place to instrument employee to store all non-refrigerated perishable food in sealed containers for pest management and providing garbage cans with a lid or hands-free operation.

4. Conclusion

Promoting workplace health is a complex challenge that requires multi-layers of approaches from various disciplinary solutions in a collaborative manner to increase

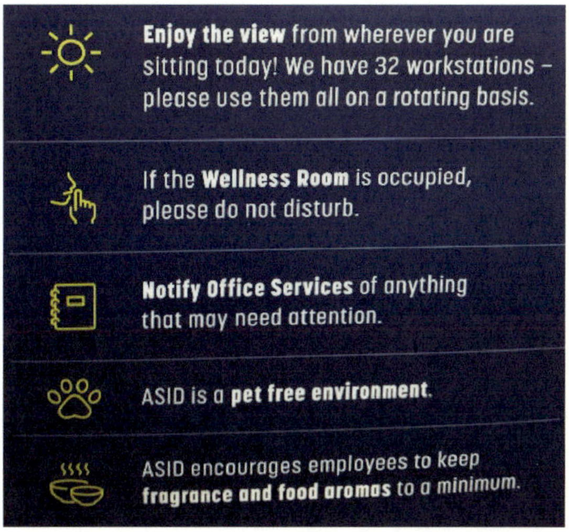

Figure 6.
ASID office etiquette list posted in the open-plan office. Photo courtesy: the American Society of Interior Designers.

physical exercise, healthy diet, mental health, and healthy lifestyle all together to change the current health epidemics. These approaches include not only technical measures but also spatial planning, design strategies, and workplace policies (**Figure 6**). The PROWELL model and its 7 KPIs of WH suggest a comprehensive approach to workplace health to address this. While 7 KPIs of WH address comprehensive topics, WH is not limited to only the strategies and specific measures discussed in this chapter. Other strategies, thus, can be further implemented to increase mental health and social well-being through various assistance programs and workplace policies to provide a balanced health in all three domains of health. Individual workplaces can also seek specific KPIs and strategies to promote a particular target domain or organizational outcomes more desirable to a workplace under specific pursuits and circumstances.

It is worth noting that while many health interventions at work have focused on health impacts at the individual level, companies are more interested in understanding the impacts of workplace health at the organizational level by exploring organizational outcome measures such as the sickness absenteeism, accident rates, work-related injuries, presenteeism, job performance, employee engagement, and retention rates [78]. Through decades-long research in public health, we have learned certain organizational outcome measures are more meaningful to the organization level decisions. For instance, sickness absence analysis tends to be less meaningful as a health measure to organizational decision-making in establishing better health promotion strategies, or motivating employees to engage in healthier behaviors, compared to other types of measures [79].

When looking for workplace health outcomes at the organizational level, it is important to note many discrepancies shown in various studies. These discrepancies are likely due to validity and reliability issues in research designs. The poorer the research methodological quality is, the higher the effects are [80, 81]. Thus, we need to be aware of possibilities of either under- or over-valuation of certain health interventions in research reports instead of trusting the impacts interpreted as exact numbers and dollar amounts. What is important is that there is a consistent discovery of the benefits of workplace health in both employee health and organizational outcomes. At the same time, it is important to develop further proper research methods and designs to yield more accurate results to narrow the discrepancies.

Acknowledgements

I would like to express my gratitude to many colleagues and experts who supported me with the development of PROWELL and writing of this book chapter. I send special thanks to Barbara Marini, James Rice, Janice Barnes, Joseph Connell, Susan Chung, and Bryan Steverson. And a heartfelt thank you goes to Matthew Schottenfeld, Assistant Director in Business of Innovative Workplace Institute, for his long time support for my work and reviewing this book chapter.

Author details

Young Lee[1,2]

1 Innovative Workplace Institute, NY, USA

2 University College London, London, UK

*Address all correspondence to: young.lee@innovativeworkplaceinstitute.org

References

[1] World Health Organization. Chronic diseases and health promotion. 2019. [Online]. Available from: https://www.who.int/chp/about/integrated_cd/en/ [Accessed: 22 January 2019]

[2] World Health Organization. Obesity and overweight key facts. 2018. [Online]. Available from: https://www.who.int/news-room/fact-sheets/detail/obesity-and-overweight. [Accessed 14 December 2018]

[3] Popkin BM, Kim S, Rusev ER, Du S, Zizza S. Measuring the full economic cost of diet, physical activity and obesity-related chronic diseases. Obesity Reviews, 2006, 20067, 271-293

[4] World Health Organization. World health report: Mental disorders affect one in four people. 2018. [Online]. Available from: https://www.who.int/whr/2001/media_centre/press_release/en/ [Accessed 14 December 2018]

[5] World Health Organization. Global Action Plan on Physical Activity 2018-2030: More Active People for A Healthier World. Geneva: World Health Organization; 2018

[6] World Health Organization. Obesity Swallows Rising Share in GDP in Europe: Up to 1% and Counting. London: WHO; 2006

[7] Chartered Institute of Personnel & Development. Absence Management. London: Chartered Institute of Personnel and Development; 2007

[8] Chartered Institute of Personnel & Development. Recruitment, Retention and Turnover. London: Chartered Institute of Personnel & Development; 2007

[9] European Commission. Improving the mental health of the population: Towards a strategy on mental health for the European Union. In: European Commission in Brussels. 2005

[10] Gallup. State of the American Workplace: Employee Engagement Insights for U.S. Business Leaders. Washington, DC: Gallup; 2013

[11] Chu C, Breucker G, Harris N. Health-promoting workplaces—International settings development. Health Promotion International. 2000;**15**(2):155-167

[12] Schulte PA, Wagner GR, Ostry A, Blanciforti LA, Cutlip RG, Krajnak KM. Work, obesity, and occupational safety and health. American Journal of Public Health. 2004;**97**(3):428-436

[13] Cooper G, Dewe P. Well-being—Absenteeism, presenteeism, costs and challenges. Occupational Medicine. 2008;**2008**(58):522-524

[14] The European Agency for Safety and Health at Work. EU-OSHA Annual Report 2008: Health and Safety in Hard Times. Luxembourg: Office for Official Publications of the European Communities; 2008

[15] World Health Organization. Constitution of the World Health Organization. World Health Organization; 1946. [Online]. Available from: http://apps.who.int/gb/bd/PDF/bd47/EN/constitution-en.pdf?ua=1 [Accessed 25 January 2019]

[16] David Lock Associates. Active Design: The Role of Master Planning, Phase 1, September 2005. London: Sport England; 2005

[17] Sport England. Active Design: Promoting Opportunities for Sport and Physical Activity through Good Design. London: Sport England; 2007

[18] Biederman J, Petty C, Clarke A, Lomedico A, Faraone S. Predictors of persistent ADHD: An 11-year follow-up study. Journal of Psychiatric Research. 2011;**45**(2):150-155

[19] Ikemi M. The effects of mystery on preference for residential façades. Journal of Environmental Psychology. 2005;**25**(2):167-173

[20] Browning W, Ryan C, Clancy J. 14 Patterns of Biophilic Design. New York: Terrapin Bright Green; 2014

[21] Grimstvedt ME, Kerr J, Oswalt SB, Fogt DL, Vargas-Tonsing TM, Yin Z. Using signage to promote stair use on a university campus in hidden and visible stairwells. Journal of Physical Activity & Health. 2010;**7**(2):232-238

[22] World Health Organization. Global Recommendations on Physical Activity for Health. Switzerland: World Health Organization; 2010

[23] Schuna JM, Swift DL, Hendrick CA, Duet MT, Johnson WD, Martin CK, et al. Evaluation of a workplace treadmill desk intervention. American College of Occupational and Environmental Medicine. 2014;**56**(12):1266-1276

[24] The UK Society of Light and Lighting> LG07/15 Lighting Guide 07: Offices—LG7. London: The Chartered Institute of Building Services Engineers; 2015

[25] The US Department of Labor. Computer Workstations Etool—Workstation Environment. [Online]. Available from: https://www.osha.gov/SLTC/etools/computerworkstations/wkstation_enviro.html [Accessed: 10 February 2019]

[26] I.C. Council. 2018 IECC—International Energy Conservation Code. Washington DC, US: International Code Council; 2017

[27] U. G. B. Council. LEED ID + C: Commercial Interiors V4: Daylight. [Online]. Available from: https://www.usgbc.org/credits/commercial-interiors-retail-commercial-interiors-hospitality-commercial-interiors/v4-draf-10 [Accessed 9 February 2019]

[28] The US General Services and Administration. The Facilities Sandards for the Public Buildings Service. Washington, DC: The US General Services Administration; 2014

[29] ASHRAE. Advanced Energy Design Guide for Small to Medium Office Buildings, Atlanta, GA: The American Society of Heating, Refrigerating and Air-Conditioning Engineers, 2011

[30] British Standards Institution. BS EN 12464-1:2011 Light and Lighting—Lighting of Work Places. London: British Standards Institution; 2011

[31] Lee Y, Boatman E, Jowett S, Guenther B. REAL: The technology-enabled, engaged, and active learning classroom. International Journal of Designs for Learning. 2014;**5**(1):57-67

[32] Illuminating Engineering Society. ANSI/IES RP-1-12 American National Standard Practice for Office Lighting. New York: Illuminating Engineering Society; 2013

[33] Schumpeter JA. Capitalism, Socialism and Democracy. Oxfordshire, UK: Routledge; 1994

[34] Öhrström E, Björkman M, Rylander R. Noise annoyance with regard to neurophysiological sensitivity, subjective noise sensitivity and personality variables. Psychological Medicine. 1988;**18**(3):605-613

[35] Beheshti MH, Hajizadeh RR, Borhani Jebeli MB, Tajpoor A. The role of individual and personality traits in noise annoyance. Annals of Medical and Health Sciences Research. 2018;**8**:133-138

[36] Shepherd D, Heinonen-Guzejev M, Michael J Hautus3 MJ, Heikkilä2 K, Hautus MJ, Heikkilä K. Elucidating the relationship between noise sensitivity and personality. Noise and Health. 2015;**17**(76):165-171

[37] The US General Services and Administration. Sound Matters: How to Achieve Acoustic Comfort in the Contemporary Office. Washington, DC: The US General Services and Administration; 2011

[38] American Society for Healthcare Engineering Association. Guidelines for the Design and Construction of Hospitals and Outpatient Facilities. Illinois: The Facility Guidelines Institute; 2014

[39] The US Green Building Council. LEED v4 Building Design and Construction. Washington DC, US: The US Green Building Council; 2019

[40] Geiler and Associates. Introduction to Sound Trasmission Class. LEED User. [Online]. Available from: https://leeduser.buildinggreen.com/sites/default/files/credit_documentation/STC%20Introduction.pdf [Accessed 11 February 2019]

[41] Roper K, Payant R. The Facility Management Handbook. New York, US: AMACOM; 2014

[42] Wyon DP, Andersen I, Lundqvist GR. Spontaneous magnitude estimation of thermal discomfort during changes in the ambient temperature. Journal of Hygiene. 2009;**70**(2):203-221

[43] Kingma B, Van Marken Lichtenbelt W. Energy consumption in buildings and female thermal demand. Nature Climate Change. 2015;**5**:1054-1056

[44] Karjalainen K. Biological sex differences in thermal comfort nad use of thermostats in everyday thermal environments. Building and Environment. 2007;**4**(1594-1603):42

[45] ASHRAE. ASHRAE 55-2017: Thermal Environmental Conditions for Human Occupancy. Atlanta, Georgia, US: The American Society of Heating, Refrigerating & Air-conditioning Engineers; 2017

[46] Steinemann A. Prevalence and effects of multiple chemical sensitivities in Australia. Preventive Medicine Reports. 2018;**10**:191-194

[47] Bernard BP. Musculoskeletal Disorders and Workplace Factors: A Critical Review of Epidemiologic Evidence for Work-Related Musculoskeletal Disorders of the Neck, Upper Extremity, and Lower Back. Cincinnati: National Institute for Occupational Safety and Health; 1997

[48] National Research Council and Institute of Medicine, Musculoskeletal Disorders and the Workplace: Low Back and Upper Extremities. Panel on Musculoskeletal Disorders and the Workplace. Washington, DC: National Academy Press; 2001

[49] Joint WHO/FAO Expert Consultation. WHO Technical Report Series 916: Diet, Nutrition and the Prevention of Chronic Diseases. Geneva: The World Health Organization; 2003

[50] Willett WC, Koplan JP, Nugen R, Dusenbury C, Puska P, Gaziano TA. Prevention of chronic disease by means of diet and lifestyle changes. In: Disease Control Priorities in Developing Countries. 2nd ed. New York: Oxford University Press; 2006. pp. 833-850

[51] Salvendy G. Handbook of Human Factors and Ergonomics. John Wiley & Sons; 2012

[52] Nonaka I, Takeuchi H. The Knowldge-Creating Company: How

Japanese Companies Create the Dynamics of Innovation. London: Oxford University Press; 1995

[53] Kim S, Park YA, Niu Q. Micro-break activities at work to recover from daily work demands. Journal of Organizational Behavior. 2017;**38**(1):28-44

[54] Schwartz T, Porath C. Why you hate work. The Energy Project. New York Times; 30 May 2014

[55] Alavi M, Leidner DE. Knowledge management and knowledge management systems: Conceptual foundations and research issues. MIS Quarterly. 2001;**25**(1):107-136

[56] Ahmed PK. Culture and climate for innovation. European Journal of Innovation Management. 1998;**1**(1):30-43

[57] Banbury SP, Berry DC. Office noise and employee concentration: Identifying causes of disruption and potential improvements. Ergonomics. 2005;**48**(1):25-37

[58] Cacioppo JT, Patrick W. Loneliness: Human Nature and the Need for Social Connection. New York, US: W. W. Norton & Company; 2008

[59] Valtorta NK, Kanaan M, Gilbody S, Ronzi S, Hanratty B. Loneliness and social isolation as risk factors for coronary heart disease and stroke: Systematic review and meta-analysis of longitudianl observational studies. Heart. 2016;**102**:1009-1016

[60] Seppälä E, King M. Burnout at work isn't just about exhaustion: It's also about loneliness. Harvard Business Review. 29 June 2017

[61] Davis MC, Leach DJ, Clegg CW. The physical environment of the office: Contemporary and emerging issues. In: International Review of Industrial and Organizational Psychology. Chichester: Wiley; 2011. pp. 193-235

[62] Waber B. People Analytics: How Social Sensing Technology Will Transform Business and What it Tells us about the Future of Work. Upper Saddle River, New Jersey, US: Financial Times/ Prentice Hall; 2013

[63] DeCanio SJ, Watkins WE. Information processing and organizational structure. Journal of Economic Behavior & Organization. 1998, 1998;**36**:275-294

[64] Maslow AH. A theory of human motivation. Psychological Review. 1943;**50**(4):370-396

[65] Home R, Hunziker M, Bauer N. Psychosocial outcomes as motivations for visiting nearby urban green spaces. Leisure Sciences. 2012;**34**(4):350-365

[66] Thompson CW, Roe J, Aspinall P, Mitchell R, Clow A, Miller D. More green space is linked to less stress in deprived communities: Evidence from salivary cortisol patterns. Landscape and Urban Planning. 2012;**105**(3):221-229

[67] Berman MG, Jonides J, Kaplan S. The cognitive benefits of interacting with nature. Psychology Science. 2008;**19**(12):1207-1212

[68] Smarr BL, Jennings KJ, Driscoll JR, Kriegsfeld LJ. A time to remember: The role of circadian clocks in learning and memory. Behavioral Neuroscience. 2014;**128**(3):283-303

[69] Bailey SM, Udoh US, Young ME. Circadian regulation of metabolism. Journal of Endocrinology. 2014;**222**(2):R75-R96

[70] Kim TW, Jeong JH, Hong SC. The impact of sleep and circadian disturbance on hormones and metabolism. International Journal of Endocrinology. 2015;**2015**:1-10

[71] Steverson B. Circadian Light for your Health. Chicago: The U.S. General Services Administration; 2017

[72] Bungay H, Munn-Giddings C, Boyce M, Wilson C. The Value of the Arts in Therapeutic and Clinical Interventions. East Anglia, UK: Anglia Ruskin University; 2014

[73] Moultrie J, Nilsson M, Dissel M, Haner U, Janssen S, Van der Lugt R. Innovation spaces: Towards a framework for understanding the role of the physical environment in innovation. Creativity and Innovation Management. 2007;**16**(1):53-65

[74] Wells MM, Thelen L, Ruark J. Workspace personalization and organizational culture: Does your workspace reflect you or your company? Environment and Behavior. 2007;**39**(5):616-634

[75] Dinc P. Gender difference in private offices: A holistic approach for assessing satisfaction and personalization. Journal of Environmental Psychology. 2009;**29**(1):53-62

[76] Bluyssen PM. Towards new methods and ways to create healthy and comfortable buildings. Building and Environment. 2010;**45**(4):808-818

[77] Klitzman S, Stellman JM. The impact of the physical environment on the psychological well-being of office workers. Social Science & Medicine. 1989;**29**(6):733-742

[78] Pereira MJ, Coombes BK, Comans TA, Johnston V. The impact of onsite workplace health-enhancing physical activity interventions on worker productivity: A systematic review. Occupational and Environmental Medicine. 2015;**72**(6):401-412

[79] Hubera M, Lechnerb M, Wunschc C. Workplace health promotion and labour market performance of employees. Journal of Health Economics. 2015;**43**(2015):170-189

[80] Palmer KT, Harris EC, Linaker C, Barker M, Lawrence W, Cooper C, et al. Effectiveness of community and workplace-based interventions to manage musculoskeletal-related sickness absence and job loss: A systematic review. Rheumatology (Oxford). 2012;**51**(2):230-242

[81] Rongen A, Robroek SJ, Van Lenthe FJ, Burdorf A. Workplace health promotion: A meta-analysis of effectiveness. American Journal of Preventive Medicine. 2013;**4**(44):406-415

Chapter 4

Trials for Health Promotion by Indoor Environment Modifications

Suni Lee, Naoko Kumagao-Takei, Kei Yoshitome, Nagisa Sada, Yasumitsu Nishimura and Takemi Otsuki

Abstract

Two attempts to address health issues by indoor environment modifications are introduced. One approach involves enhancement of natural killer cell activity by negatively charged particle dominant indoor air conditions (NCPDIAC) resulting from extra-porous charcoal paint and application of an electric voltage on the wall surface to adsorb positively charged small particles (approx. 20 nm in diameter). This apparatus creates relatively continuous NCPDIAC. The other approach involves prevention of pollen allergy symptoms by a cloth containing specific ore powder (CCSNOP). With the former approach, we engaged in short-term (2.5 hour), middle-term (2 W night stay), and long-term (3 M) stays under NCPDIAC and found that IL-2 levels increased under short-term stays and that natural killer cell activity was enhanced in middle- and long-term stay experiments. Implementation of this strategy may partially prevent the occurrence of cancers and viral-mediated diseases. With the latter approach, a 1-hour stay within a CCSNOP room resulted in improvement of symptoms in patients with pollen allergies in addition to a reduction in bad moods caused by any remaining symptoms. In both cases, longer-term experiments should be performed in an effort to confirm and delineate the biological effects of these indoor environment modifications on human health problems.

Keywords: indoor air, negatively charged particle, natural killer cell activity, pollen allergy

1. Introduction

The nature of indoor environments is important in terms of human health. One typical illness related to indoor environments is referred to as sick building syndrome (SBS) and is mainly caused by volatile organic compounds (VOCs) [1–6]. Patients with SBS suffer from a variety of nonspecific symptoms such as irritation of the eyes, nose, or throat. Additionally, certain neurotoxic or general health problems present as skin irritations, with nonspecific hypersensitivity reactions including infectious diseases and odor and taste sensations [1–6]. These patients also suffer from general fatigue. Many efforts have been directed at exploring the etiology of SBS, such as attempts to identify patients by certain genetic or psychological features [1–6]. However, efforts to unambiguously account for the occurrence of this syndrome and strategies to prevent its occurrence by avoiding exposure to specific chemicals remain unclear [1–6].

On the other hand, various attempts have been made to improve human health by modifying indoor air environments. For example, efforts have been devoted to increasing air tightness and thermal insulation properties to avoid marked changes between rooms [7–10]. Additionally, antifungal measures are also important since a variety of fungi can cause health problems such as hypersensitivity pneumonitis and other allergic diseases [11, 12].

In this chapter, two approaches to improve human health by indoor environment modifications are introduced. One approach involves "enhancement of natural killer cell activity caused by negatively charged particle dominant indoor air conditions," while the other involves "improvement of pollen allergy symptoms by cloth containing specific ore powder."

2. Enhancement of natural killer cell activity caused by negatively charged particle dominant indoor air conditions

We previously described the construction of negatively charged particle dominant indoor air conditions (NCPDIAC) [13]. Thereafter, the effects of NCPDIAC on human health have been monitored under short- (2.5 hours), middle- (2-week night stay), and long-term (3-month period of continuous residence) stays [13].

As shown in **Figure 1A**, the NCPDIAC apparatus consists of two components. One comprises the use of extra-porous charcoal paint [13]. This mainly induces dehumidification and deodorization. The important component that enhances natural killer (NK) cell activity comprises an electric voltage on the wall surface (72–100 V) [14, 15]. This electric voltage creates a negative charge on the wall surface [13]. Thus, positively charged particles (approx. 20 nm in diameter) are adsorbed onto the wall surface, and negatively charged particles remain at random in the rooms. As a result, this device creates NCPDIAC.

As shown in **Figure 1B–D** panels, three different types of experiments were performed sequentially. In the first, 2.5 hour short-stay experiments were performed. At that time, three control rooms and three NCPDIAC rooms were built in the wide sub-underground laboratory in the Comprehensive Housing R&D Institute, Sekisui House Ltd., at Kyoto Prefecture, Japan. The control and NCPDIAC rooms were 9.1 m^2 (floor area) and 22.8 m^3 (air volume). As described in **Figure 1B**, 60 volunteers occupied the control or NCPDIAC rooms for a 2.5 hour stay (short-term). There were no differences in temperature, humidity, or concentration of VOCs between control and NCPDIAC rooms during the experimental period. However, the concentration of charged particles differed [13]. As shown in **Figure 1B**, the concentration of positively charged particles in control rooms was slightly but significantly higher than that of negatively charged particles. It was considered that this difference may represent ordinary conditions. However, in NCPDIAC rooms, the concentration of positively charged particles decreased (by adsorption onto the wall surface), whereas that of negatively charged particles remained unchanged. Thus, the difference in concentration between positively and negatively charged particles was significantly greater (approximately 500–800 particles/cm^3 room air) [13].

A second series of experiments was performed comprising 2-week night stays [14]. Fifteen volunteers (all Japanese males) occupied dormitory or Sekisui House rooms for 3 months. At the mid-month period, participants moved into control rooms (no NCPDIAC apparatus), and all participants were now subject to the same conditions at this point in time [14]. However, participants did not know which of the rooms (control or NCPDIAC) they had initially occupied. Participants stayed

DOI: http://dx.doi.org/10.5772/intechopen.84153

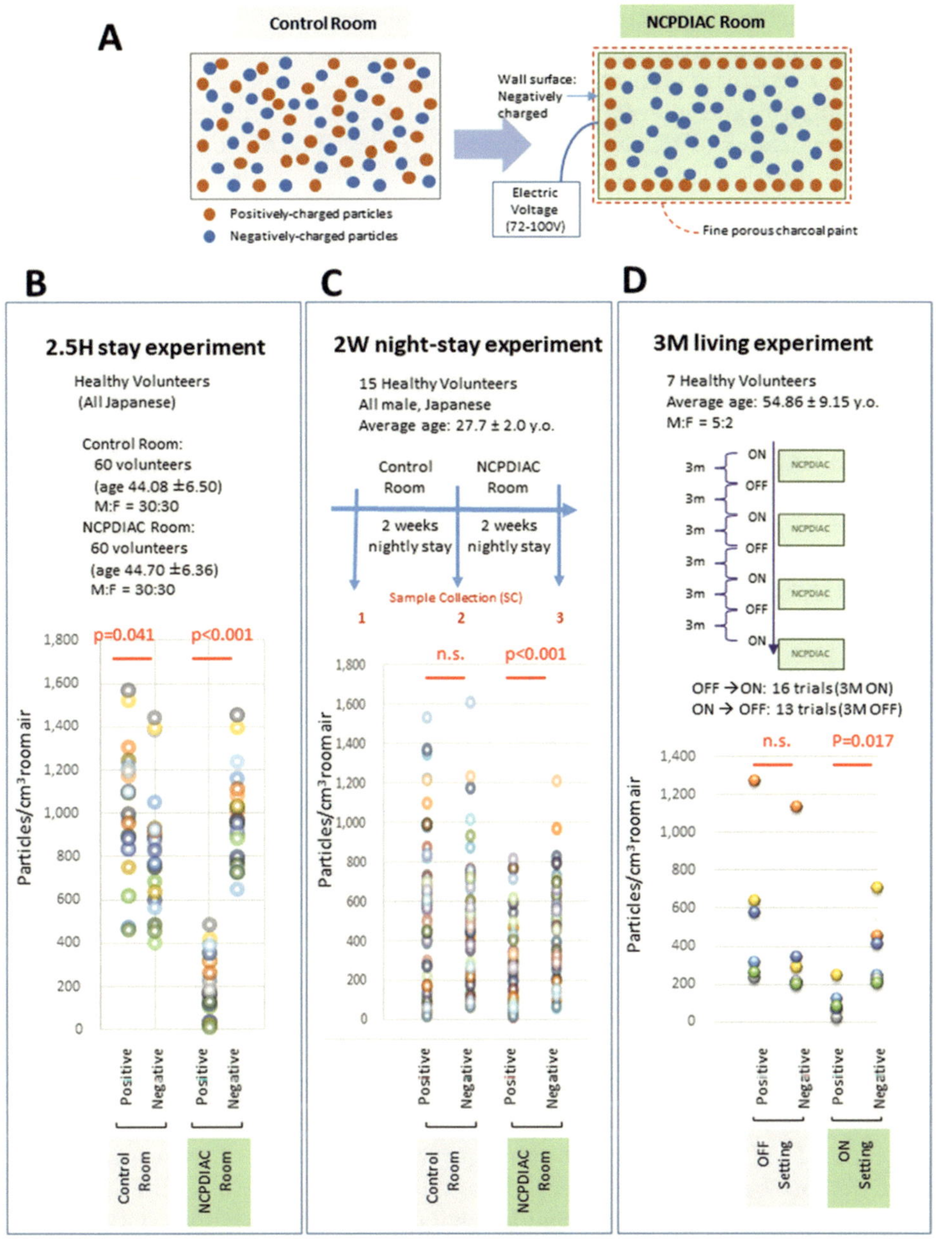

Figure 1.

(A) The mechanism employed to produce negatively charged air dominant indoor air conditions (NCODIAC). Fine porous charcoal paints were painted on the walls, and an electric voltage (72–100 V) was applied to the walls. Hence, wall surfaces became electrically negatively charged. As a result, positively charged particles were adsorbed onto the surface of walls. Therefore NCPDIAC was formed continuously and relatively. (B) 2.5 hour stay experiments were performed with 60 healthy volunteers. In this experiment, control and NCPDIAC rooms were made available. The concentration of positively and negatively charged particles is shown. In control rooms, a slight increase in positively charged particles was found; however, in NCPDIAC rooms the concentration of positively charged particles decreased. Although the concentration of negatively charged particles remained unchanged compared with control rooms, differences resulting from NCPDIAC were significant. (C) Two week night-stay experiments were performed using 15 healthy volunteers. Control and NCPDIAC rooms were made available within a dormitory. Positively and negatively charged particles were measured in control and NCPDIAC rooms on three separate occasions. As a result, NCPDIAC rooms showed a decrease in positively charged particles and a relatively dominant condition of negatively charged particles. (D) Thereafter, 3-month stay experiments were performed with participants in their own home. There were 16 trials comprising OFF (including precondition) to ON periods and 13 trials comprising ON to OFF periods. It was found that during the OFF period, there was no difference between the concentrations of positively and negatively charged particles. However, during the ON period, the concentration of negatively charged particles was significantly higher compared with that of positively charged particles (the concentration of which had decreased).

for 2 weeks during the evenings (night stay) while receiving employee training during the daytime and then moved into NCPDIAC rooms for the following 2 weeks. Thereafter, participants returned to the rooms they had initially occupied. As with the 2.5 hour stay experiments, there were no differences in temperature, humidity, or concentration of VOCs between control and NCPDIAC rooms during the experimental period [14]. However, the concentration of charged particles differed. As shown **Figure 1C**, the concentration of positively charged particles decreased in the NCPDIAC rooms, and there was a significant difference between positively and negatively charged particle concentrations [14].

A third series of experiments was performed comprising 3-month stays [15]. The NCPDIAC apparatus was installed in the bedrooms or living rooms of newly built or renovated homes or condominiums. Seven volunteers participated. Volunteers would switch the NCPDIAC apparatus ON or OFF themselves every 3 months [15]. A total of 16 OFF to ON trials (ON trials) and 13 ON to OFF trials (OFF trials) were performed [15]. When the switch was OFF, there were no differences in concentration between positively and negatively charged particles, whereas when the switch was ON, the concentration of positively charged particles decreased, and there was a significant difference between positively and negatively charged particle concentrations [15], as shown in **Figure 1D**.

3. Biological effects of NCPDIAC

Prior to and following each of the stay periods (2.5 hour, 2-week night, and 3-month ON/OFF trials of NCPDIAC), biological measurements were performed [13–15].

In the 2.5 hour stay experiments, various general medical checks were performed such as peripheral blood counts, liver and kidney functions, lipids, minerals as well as immunoglobulin, and stress biomarkers (including serum cortisol, salivary immunoglobulin (Ig) A, chromogranin, and amylase), as well as a questionnaire given for mood, various autonomic nerve conditions (heart rate, blood pressure, body sway, and Flicker test), and blood viscosity. Cytokinesrelated to balance between T helper (Th) 1 and Th2 such as interferon (IFN)-γ, tumor necrosis factor (TNF)-α, and interleukin (IL)-2 for Th1, and IL-4, IL-6, and IL-10 for Th2 were also measured.. All variables were recorded as [post-stay]–[pre-stay] values. These values were compared between stays in control and NCPDIAC rooms [13]. As shown in **Figure 2A**, the most significant difference was found in the level of IL-2 [13]. Although the changes were very small, there was a significant difference between stays in control and NCPDIAC rooms. Additionally, statistical significance was observed for changes in IL-4 (higher in NCPDIAC rooms) and fluctuations in heart rate (lower in NCPDIAC rooms). However, the most noticeable difference was in IL-2 [13].

Two-week night-stay experiments were then performed. For this series of experiments, additional items that were checked included NK activity, urine 17-hydroxycorticosteroid (OHCS) as a stress marker, and deoxyguanosine (8-OHdG) as an oxidative stress marker, since it was considered that these factors may change during the 2-week period, even though they remain unchanged in the short term [14]. The only difference observed for the various biomarkers examined prior to and following occupation of the NCPDIAC rooms was in NK cell activity. As shown in **Figure 2B**, relative NK cell activity tended to increase with sample collection (SC) 3 (after a 2-week night stay in NCPDIAC rooms) and SC1 as 1.0, while a significant increase was observed with SC3 and SC2 as 1.0. This was considerable at this point, as the short-term experiments revealed a significant increase in IL-2 [14]. However, IL-2

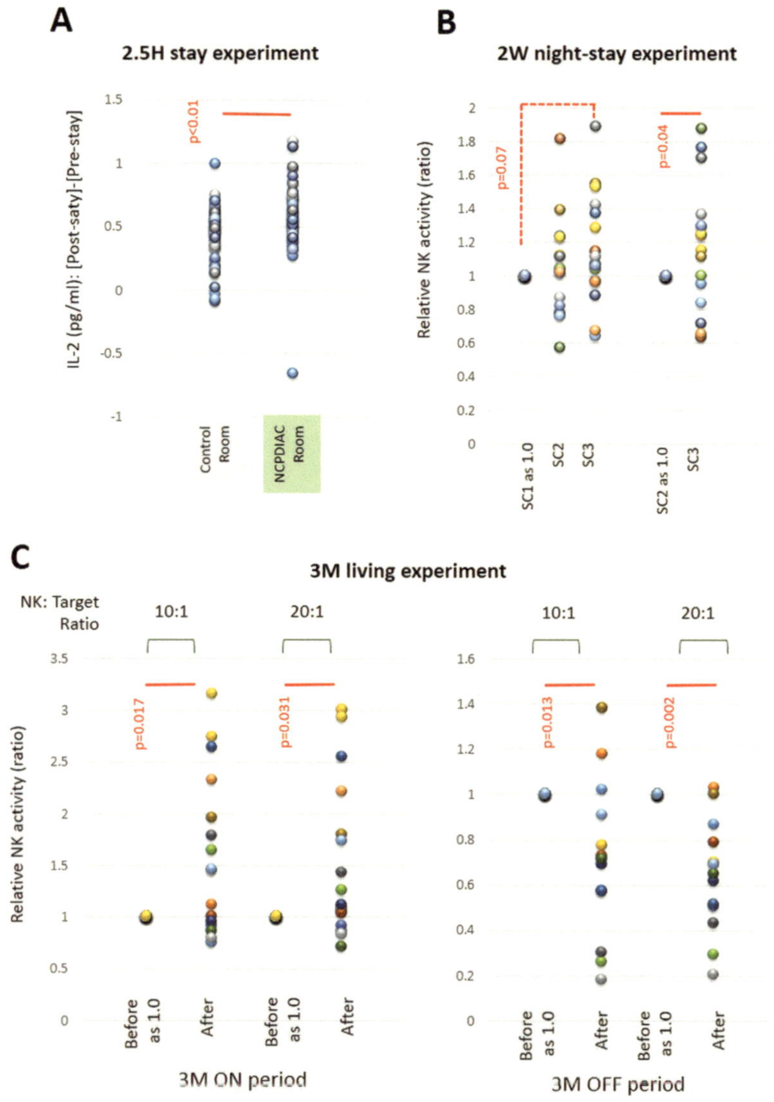

Figure 2.
(A) Of the various biological tests that were employed for the 2.5 hour stay experiments, the greatest change observed in the [post-stay]–[pre-stay] values was in the concentration of serum IL-2. Stays in the NCPDIAC rooms resulted in significant increases in IL-2. (B) During the 2 W night-stay experiments, the relative NK cell activity increased significantly after stays in the NCPDIAC rooms. The relative NK cell activity at SC (sample collection) 3 (after the 2 W night stay in NCPDIAC rooms) compared with SC1 (at the time of leaving control rooms) tended to increase, and when compared with SC2 (just before entering NCPDIAC rooms, after 2 W night stay in control rooms), there was a significant increase in NK cell activity. (C) In the 3 M stay experiments during the 3 M ON period (left panel), the relative NK cell activity increased following stays when activities were measured using 10:1 and 20:1 ratios (NK cells versus target cancer cells in vitro). Additionally, during the OFF period, the relative NK cell activity decreased significantly under 10:1 and 20:1 conditions.

levels were considered to return to basal levels after participants had ceased to occupy NCPDIAC rooms. The slight but significant increase in IL-2 levels during the night-time for the 2-week period may lead to activation of NK cells [14].

The third series of experiments involved participants living in their own homes [15]. Participants had the NCPDIAC apparatus in their bedrooms or living rooms. This apparatus was switched ON or OFF every 3 months by the volunteers themselves. Biological measurements were performed prior to the ON period (including the initial pre-time) and 3 months later (just prior to the OFF period) and then again just prior to the ON period (3 months later). As shown in **Figure 2C**, relative NK cell activity increased during the ON period and decreased during the OFF period [15]. The left panel of **Figure 2C** shows a 3-month ON period, with both NK cell and target cell ratios (10:1 and 20:1, respectively) shown [15]. In both cases, after a 3-month ON period, the relative NK cell activity increased [15]. On the other hand, as shown in the right panel of **Figure 2C**, during the OFF period, even with cell ratios of 10:1 and 20:1, the NK cell activity decreased [15].

NK cell activity is known to be correlated with the presence of cancer cells in humans and with viral infection of cells. Thus, indoor environments that yield enhancements in NK cell activity may somehow lead to the prevention of cancers in occupants and improvement of symptoms associated with viral-mediated illnesses such as influenza.

4. Other findings regarding NCPDIAC

In addition to the aforementioned results, in vitro studies were performed [16]. Peripheral blood mononuclear cells from healthy volunteers were cultured in a standard CO_2 incubator or with forced influx using an adsorbed negatively charged air incubator. With the latter (experimental) incubator, the concentration of negatively charged particles was higher, at approx. 3000 particles/cm^3 air in the incubator [16].

It was found that NK cell activity was enhanced in the experimental incubator. Moreover, CD25 expression in CD4 T cells and IFN-γ production in supernatants were also enhanced following application of the experimental incubator. These findings also support the notion that NCPDIAC may stimulate the immune status, but not quite pathologically [16]. Then, to demonstrate a representative phenotype, NK cell activity was found to be enhanced as shown in **Figure 2C**.

Furthermore, in the 3-month stay experiments, serum amyloid A (SAA) levels decreased significantly during the ON period [17]. Although the exact meaning of this finding is unclear, SAA is considered to be an acute reactive protein such as C-reactive protein (CRP) [18]. Additionally, the increase in high sensitivity CRP is considered to represent a predictive biomarker of atherosclerosis, and, if SAA levels also fluctuate similar to CRP, NCPDIAC may prevent the progression of atherosclerosis [17].

At present, our efforts are being directed at monitoring participants engaged in living under NCPDIAC for greater than 1 year.

5. Improvement in pollen allergy symptoms by cloth containing specific ore powder

Next, a trial aimed at preventing pollen allergies by indoor environment modifications was introduced [18].

We used specific natural ore powder (SNOP) in the form of a cloth containing this powder (CCSNOP). Details of the experiments have been reported previously. SNOP was obtained near Mount Aso, one of the biggest volcanoes in Japan (Kumamoto Prefecture), and is known to release far-infrared rays [18]. Chemical analyses revealed no specific characteristics of the natural ore found surrounding the Mount Aso area and mainly comprise SiO_2, Al_2O_3, Fe_2O_3, Na_2O, and other small chemical compounds. The Cosmic Garden Co. Ltd., a custom-home building company located in Okayama City, Japan, is the company from which our materials are sourced [18]. Their products and homes contain features of "2 × 4" construction methods for aseismic capacity, the avoidance of chemical substrates and smells by well-sealed and super-insulated areas, and enhanced durability. Additionally, this company has employed SNOP within wall materials. Since clients utilizing these products have provided anecdotal accounts of improvements in symptoms related to pollen allergies, we attempted to investigate these phenomena and provide empirical evidence for any health benefits [18].

As shown in **Figure 3**, 20 pollen allergy patients were recruited in April (the season for pollen allergies in Japan) of 2014. The average age of participants was 36.9 ± 11.2 years and comprised of 11 males and 9 females [18]. Over a 2-week period, participants occupied rooms covered by CCSNOP or by non-woven cloth (NWC), which did not contain the ore powder, for 1 hour. During this 1-hour period, windows were opened for 1–2 minutes every 15 minutes to expose the rooms

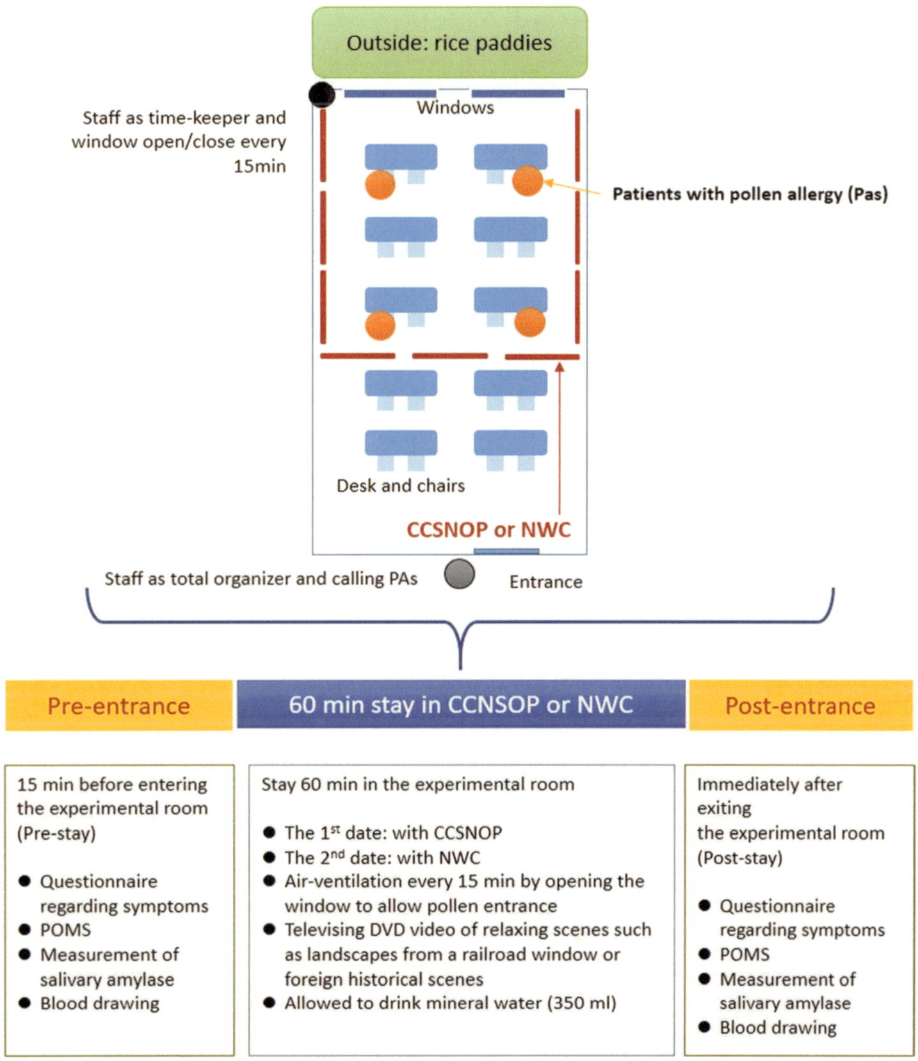

Figure 3.
Twenty patients with pollen allergies occupied rooms covered by CCSNOP or NWC in the spring (season for pollen allergies in Japan) of 2014. During the 1-hour stay in the CCSNOP or NWC rooms, windows were opened for 1–2 minutes every 15 minutes to expose the rooms to pollen. Participants were subjected to various tests as shown in the lower panel. Prior to and following entry into the rooms, participants completed a questionnaire related to symptoms and mood [profile of mood status (POMS)], as well as tests for salivary amylase as a stress biomarker, and blood drawing. The questionnaire was evaluated by the chi-square test, and CCSNOP and NWC stays were compared. Other biological markers were also compared and represented as [post-entrance]–[pre-entrance] values.

to pollen [18]. Prior to and following occupation of the rooms, participants completed a questionnaire regarding symptoms and mood (a POMS (profiles of mood status) questionnaire was employed), and a variety of blood tests were performed including peripheral blood counts, liver and kidney functions, minerals, specific Ig E for various allergens, and cytokines. The measured items were represented as [post-admission] – [pre-admission] [18].

The results of the questionnaire of symptoms and mood are shown in **Figure 4**. In the upper panel, it can be seen that nasal obstruction, lacrimation, and difficulty of daily life improved in participants occupying the CCSNOP rooms compared with those occupying the NWC rooms. However, eye redness became worse in participants occupying the CCSNOP rooms. In the lower panel, moods for "tension and anxiety," "depression," and "fatigue" improved in participants occupying the CCSNOP rooms compared with those occupying the NWC rooms. Additionally, total mood disturbance also improved. All of these statistical analyses employed the chi-square test [19].

Total Ig E was found to increase slightly but significantly in participants occupying the CCSNOP rooms compared with those occupying the NWC rooms. Moreover, the percentage of eosinophils also increased in participants occupying the CCSNOP rooms. The reason for the changes in IgE and eosinophils remains unclear. Our experiment comprised just a 1-hour stay. Thus, although symptoms and moods may change rapidly, IgE and eosinophils changes may require more time [19].

	CCSNOP		NWC		
Symptom	Improved	No change or getting worse	Improved	No change or getting worse	P value (X^2 test)
Sneezing	14	2	13	5	0.271
Nasal juice	12	6	10	6	0.464
Nasal obstruction	12	6	9	8	* 0.003
Eye redness	4	7	7	7	* 0.010
Lacrimation	7	4	4	11	* <0.001
Difficulty of daily life	11	5	5	7	* 0.021

Statistically significant differences were found for items "Sneezing", LAcrimation" and "Difficulty of daily life". These items indicated that associated symptoms improved in CCSNOP compared to NWC. However, item "Eye redness" showed an improvement that was significantly higher in NWC than in CCSNOP.

	CCSNOP		NWC		
Score Change [Post-Pre] x number of HVs	Down	Up	Down	Up	P value (X^2 test)
Tension & Anxiety	-53	0	-19	4	* <0.001
Depression	-23	2	-8	4	* 0.002
Anger & Hostility	-11	3	-8	1	0.440
Vigor	-29	6	-30	10	0.274
Fatigue	-40	9	-19	10	* 0.010
Confusion	-21	8	-23	3	0.050
Total Mood Disturbance (TMD)	-109	21	-75	39	* <0.001

Moods for "Tension & Anxiety", "Depression" and "Fatigue" decreased significantly when Health Volunteers (HVs) stayed in CCSNOP when compared to NWC. In addition, TMD decreased significantly when HVs stayed in CCSNOP compared to NWC.

Figure 4.
Results of the symptom questionnaire and POMS. Statistical data were analyzed by the chi-square test.

In future experiments we intend to make available bedrooms that can be utilized by participants with pollen allergy during spring (season of pollen allergies) for at least 2 weeks in an effort to further investigate changes in symptoms, moods, and biological factors.

6. Conclusion

We have presented the outcome of two different strategies aimed at improving human health by modification of indoor environments. A consideration of several factors is required to maintain human health and minimize physical and mental disturbances while indoors, including the avoidance of chemical substances such as VOCs and other chemicals. Future investigations will involve participants occupying variously modified rooms (including NCPDIAC and SNCBOP) for longer periods of time. These studies should contribute toward the creation of healthy indoor environments for human habitation.

Acknowledgements

All experiments presented in this chapter were approved by the Ethic Committee of the Kawasaki Medical School, Kurashiki, Japan, and written informed consent was obtained from all participants.

Conflicts of interest

For the NCPDIAC experiments, the Department of Hygiene, Kawasaki Medical School, obtained research funding from Sekisui House Ltd., Osaka, Japan, and from Yamada SXL Home (now known as Yamada Home) Co. Ltd., Takasaki, Japan. Additionally, the extra-porous charcoal paints were provided by Artech Kohboyu, Co. Ltd., Omura, Nagasaki, Japan. For the CCSNOP experiments, the Cosmic Garden Co. Ltd. provided the CCSNOP apparatus.

Author details

Suni Lee, Naoko Kumagao-Takei, Kei Yoshitome, Nagisa Sada, Yasumitsu Nishimura and Takemi Otsuki*
Department of Hygiene, Kawasaki Medical School, Kurashiki, Okayama Prefecture, Japan

*Address all correspondence to: takemi@med.kawasaki-m.ac.jp

References

[1] Stolwijk JA. Sick-building syndrome. Environmental Health Perspectives. 1991;**95**:99-100

[2] Lyles WB, Greve KW, Bauer RM, Ware MR, Schramke CJ, Crouch J, et al. Sick building syndrome. Southern Medical Journal. 1991;**84**(1):65-71, 76

[3] Laumbach RJ, Kipen HM. Bioaerosols and sick building syndrome: Particles, inflammation, and allergy. Current Opinion in Allergy and Clinical Immunology. 2005;**5**(2):135-139

[4] Burge PS. Sick building syndrome. Occupational and Environmental Medicine. 2004;**61**(2):185-190

[5] Israeli E, Pardo A. The sick building syndrome as a part of the autoimmune (auto-inflammatory) syndrome induced by adjuvants. Modern Rheumatology. 2011;**21**(3):235-239. DOI: 10.1007/s10165-010-0380-9

[6] Norbäck D. An update on sick building syndrome. Current Opinion in Allergy and Clinical Immunology. 2009;**9**(1):55-59. DOI: 10.1097/ACI.0b013e32831f8f08

[7] Carrer P, Wolkoff P. Assessment of indoor air quality problems in office-like environments: Role of occupational health services. International Journal of Environmental Research and Public Health. 2018;**15**(4):E741. DOI: 10.3390/ijerph15040741

[8] Le Cann P, Paulus H, Glorennec P, Le Bot B, Frain S, Gangneux JP. Home environmental interventions for the prevention or control of allergic and respiratory diseases: What really works. The Journal of Allergy and Clinical Immunology. In Practice. 2017;**5**(1):66-79. DOI: 10.1016/j.jaip.2016.07.011

[9] Charpin D, Baden R, Bex V, Bladt S, Charpin-Kadouch C, Keimeul C, et al. Environmental home inspection services in Western Europe. Environmental Health and Preventive Medicine. 2011;**16**(2):73-79. DOI: 10.1007/s12199-010-0171-0

[10] Wolkoff P. Indoor air humidity, air quality, and health—An overview. International Journal of Hygiene and Environmental Health. 2018;**221**(3):376-390. DOI: 10.1016/j.ijheh.2018.01.015

[11] Riario Sforza GG, Marinou A. Hypersensitivity pneumonitis: A complex lung disease. Clinical and Molecular Allergy. 2017;**15**:6. DOI: 10.1186/s12948-017-0062-7

[12] Spagnolo P, Rossi G, Cavazza A, Bonifazi M, Paladini I, Bonella F, et al. Hypersensitivity pneumonitis: A comprehensive review. Journal of Investigational Allergology & Clinical Immunology. 2015;**25**(4):237-250

[13] Takahashi K, Otsuki T, Mase A, Kawado T, Kotani M, Ami K, et al. Negatively-charged air conditions and responses of the human psycho-neuro-endocrino-immune network. Environment International. 2008;**34**(6):765-772. DOI: 10.1016/j.envint.2008.01.003

[14] Takahashi K, Otsuki T, Mase A, Kawado T, Kotani M, Nishimura Y, et al. Two weeks of permanence in negatively-charged air conditions causes alteration of natural killer cell function. International Journal of Immunopathology and Pharmacology. 2009;**22**(2):333-342

[15] Nishimura Y, Takahashi K, Mase A, Kotani M, Ami K, Maeda M, et al. Enhancement of NK cell cytotoxicity induced by long-term living in negatively charged-particle dominant indoor air-conditions. PLoS One. 2015;**10**(7):e0132373. DOI: 10.1371/journal.pone.0132373

[16] Nishimura Y, Takahashi K, Mase A, Kotani M, Ami K, Maeda M, et al. Exposure to negatively charged-particle dominant air-conditions on human lymphocytes in vitro activates immunological responses. Immunobiology. 2015;**220**(12): 1359-1368. DOI: 10.1016/j.imbio.2015.07.006

[17] Lee S, Yamamoto S, Nishimura Y, Matsuzaki H, Yoshitome K, Hatayama T, et al. Decrease in serum amyloid A protein levels following 3-month stays in negatively charged particle-dominant indoor air conditions. Biomedical and Environmental Sciences. 2018;**31**(5):335-342. DOI: 10.3967/bes2018.044

[18] Pepys MB, Baltz ML. Acute phase proteins with special reference to C-reactive protein and related proteins (pentaxins) and serum amyloid A protein. Advances in Immunology. 1983;**34**:141-212

[19] Lee S, Okamoto H, Yamamoto S, Hatayama T, Matsuzaki H, Kumagai-Takei N, et al. Biological effects of cloth containing specific ore powder in patients with pollen allergy. Biomedical and Environmental Sciences. 2016;**29**(8):563-573

Chapter 5

Indoor Air Pollutants and the Future Perspectives for Living Space Design

Igor Cretescu, Dorina Nicolina Isopescu, Doina Lutic and Gabriela Soreanu

Abstract

This study presents an overview on the indoor air pollutants and their implications in the living space design-related strategy implementation. Not only the buildings but also the cabins of diverse traveling vehicles (busses, trains, cars, spacecrafts, submarines, etc.) are envisaged in this regard. Overall, in the smart eco-efficient built environment, such indoor spaces should ensure an adequate indoor air quality (IAQ), along with accomplishing the performance for other key components such as durability, energy saving, aesthetical architecture, etc. General aspects on indoor air quality and indoor air pollution, IAQ monitoring, and remediation strategies, as well as the main types of indoor pollutants and their effects upon human health, are highlighted.

Keywords: monitoring methodology, indoor air composition, human health, indoor air quality, smart building, vehicle cabins, spacecrafts

1. Introduction

The issues of indoor air pollutants present in living spaces determine the consideration of indoor air quality (IAQ) as a priority within the environmental programs. On the other hand, the assessment of the IAQ is a complex task, taking into consideration that there are hundreds of pollutants generated by many different sources that are detected in indoor environments, depending on the particularity of each indoor space [1].

In this study, the indoor space is defined as the space for leaving, working, traveling, or other activity of people inside the buildings as well as the space inside the cabin of a traveling vehicle as trains, busses, air and spacecrafts, submarines, etc. The inhabitants in buildings or temporary users of indoor spaces are looking for a healthy indoor environment in terms of health state and well-being, based on comfort (usually ensured by the temperature and humidity control in the indoor space). However, from the point of view of the one responsible in health and environment quality, the chemical content of the indoor air, even in terms of ppm levels for some kinds of compounds, should be precisely measured and well-known. The indoor air should be free of several contaminants and odors (or limit admissible levels should be defined), including the exhaust respiratory gases, such as carbon dioxide (CO_2)

and even nitrogen oxides. Also, the particle content, in terms of PM2.5 and PM10, should be strictly monitored, because of their physiological complex and often dangerous effects.

Particularly, urban people spend most of the time indoors, which can suggest a direct relationship between IAQ and public health [1]. Moreover, IAQ applies also to other specific environments such as spacecraft cabins, which are sealed environments where the astronauts spend 100% of time during the period of their mission. Carbon dioxide can reach 7000 ppm in such environments [2]. In addition, the following compounds can be detected in spacecraft cabins (e.g., as based on NASA-Mir Program and International Space Station's atmosphere) as traces (usually from fraction to few ppm level): alcohols, siloxanes, halocarbons, aldehydes, ketones, esters, aromatics, carbon monoxide (CO), ammonia, hydrogen sulfide, sulfur dioxide, and nitrogen dioxide (NO_2) [2–6].

Overall, the indoor environment involves complex interactions between a high number of factors, being in a quite dynamic, continuous, and complex changing.

In this chapter, the only factors considered for defining the IAQ are the chemical composition and PM content, while other factors related less evidently to the air quality issues but still very important to the well-being of the inhabitants/users will be ignored (levels of lighting and sound, appropriate ergonomic conditions, work satisfaction, happiness, etc.).

When the conditions for a healthy indoor environment are not fulfilled, health problems of different seriousness levels will occur, and the most important will be presented here. Therefore, the information in this respect is equally important to be known by the specialists in the environment and health and by the large public.

2. General aspects about the indoor air pollution

The indoor air quality depends on many constantly changing parameters, requiring an interdisciplinary approach due to their complexity. These parameters are related to environment, industrial hygiene, mechanical and chemical engineering, architecture and design, and constructions, including practical advices for a better understanding of the IAQ issues.

The industrial hygiene provides information that colligates health symptoms to indoor air contaminants, serving to the identification of their sources. The main contaminant sources in this case are the architectural components of the buildings; therefore, the building process is a critical component to pursue a complete investigation and to generate a comprehensive remediation design.

Heating, Ventilation and Air Conditioning (HVAC) systems could also be sources of contaminants and may eventually distribute the pollutants from the sources to the populated parts of the buildings. The study of the HVAC is a task for the mechanical engineering discipline, contributing to design the necessary system modifications for diminishing or, ideally, eliminating the contaminating pollutants. In some cases, the HVAC system may have a beneficial role in controlling and decreasing the indoor contaminants, being the most cost-effective way to solve this problem by simply changing the ventilation track and/or air distribution [7].

In conclusion, straightforward IAQ problems can be identified and solved by knowledgeable practitioners in any of these disciplines. The more complicated situations are best addressed by an investigation comprising all these disciplines. An approach based on a specific methodology will generate a coordinated and comprehensive investigative effort. This book chapter addresses mainly to architects, mechanical engineers, building managers, and others but not to industrial hygienists.

Lately, sick building syndrome (SBS) is a popular term very often met in a specialty literature. SBS has been defined as the situation when at least 20% of the building occupants of a construction display illness symptoms for more than 2 weeks, but the source of the illness cannot be positively identified [7]. Another term associated with indoor-generating disorders is the building-related illness (BRI), when the cause of the symptoms is known. BRI can be related to a specific building source and is characterized by a distinguishable set of symptoms. Based on these concepts related to the health of the building users, a significant issue itself, in a forthcoming step, the economic aspects should be also considered, due to the high costs of medical treatment involved as a consequence of IAQ problems. Therefore, the "sick" buildings are costly due to the decrease of productivity and high absenteeism rates. According to the Environmental Protection Agency (EPA) reports, the annual loss due to a low productivity brought by poor IAQ was $60 billion [8]. Many people continue to work/live in inappropriate conditions even while experiencing symptoms related to poor IAQ, and many do not connect the illness symptoms with the air quality of the work/living places. The productivity loss is only a fraction of the cost of sick buildings. The legal implications are an ever-increasing cause for concern. Building owners and managers are the primary defendants in the rising number of lawsuits related to IAQ, but many other professions are affected: building's architects; mechanical contractors; manufacturers, sellers, installers, and consulting engineers of building air-conditioning and heating equipment; and manufacturers, distributors, sellers, and installers of carpeting, floor tiles, adhesives, and chemical products used in office machinery manufacturers, distributors, and sellers [7].

A building could be considered as a product. Anyone in the chain comprising the leasing, design, and construction of the building chain would be liable for injuries suffered by the plaintiffs. Many governmental agencies are involved in IAQ policy-making efforts, considering building design guidelines, indoor emission standard, and product labeling requirements. For example, the EPA and the World Health Organization (WHO) have estimated that 20% of the buildings in the USA have serious IAQ problems, 40% have somewhat serious problems, and 40% have no serious problems [7].

The energy-efficient designs of the actual buildings resulted in tighter building envelopes with improved insulation and low-energy-consuming ventilation, without openable windows. Under these conditions, the indoor pollutants were not sufficiently diluted with fresh air. Here, multiple aspects as the HVAC system operations and their impact on IAQ should be taken into consideration. An increase of the indoor pollutant sources is due to the tight building design. New building materials, products, and furnishings emit significant amounts of potentially hazardous chemicals into the air [7]. Moreover, this aspect could be of a special interest for specific closed environments (spacecrafts, submarines, etc.) where polluting materials should be avoided at a larger extent.

The operations of modern office equipment, including its necessary periodic cleaning and service, are another group of products and activities being a source of indoor air contaminants. Daily current activities such as the use of cleaning/disinfecting agents and the episodic use of pesticides contribute as well to the increase of the indoor pollutants level.

Among the health problems generated by the above-mentioned chemical pollutants, headaches, burning and itching eyes, respiratory difficulties, skin irritation, nausea, and fatigue are some of the common complaints. These symptoms are often vague, but they generally become worse after a longer exposure time (a whole day) spent in the indoor spaces; they disappear when the occupant leaves this environment. A number of symptoms may occur together, without a specific identifiable reason [7].

The human factor should be also taken into consideration as a contributor to the quality of the indoor environment. The presence of people, breathing and emitting body odors, affects IAQ. Human activities such as smoking significantly alter the indoor environment. Cosmetics and personal care products are sources of contaminants. The resulting situation is an increase in contaminants circulating through the indoor environment, with insufficient fresh outside air introduced through the HVAC system aiming to dilute the contaminants [7]. The risk of contaminant accumulation due to the use of different products is much higher in closed leaving environments and should be minimized as much as possible.

A detailed discussion of maintenance activities, building materials, and building operations, which affect a building's IAQ and human health, is presented in the technical-specific literature [7–11].

Another factor contributing to the IAQ issue is a change in the whole building or a certain space function, without proper modifications in the HVAC systems. In the latest decade, there have been strong increases in the commercial space renovation, for many reasons. The "modern" office building is easier to renovate than its predecessor; therefore, renovation is often a more cost-effective solution than a new edifice. When changes in the interior layout occur or when the new equipment requires additional ventilation, the technically possible changes in the HVAC systems often cannot keep up [7].

Certain other physical parameters can relate to building occupants' perception of indoor air quality. Glare from artificial lighting and visual stress from the use of visual display terminals (VDTs) can cause headaches and eye irritation. Vibration and noise may contribute to dizziness, nausea, and general irritability. Ergonomic stressors, such as chairs having wrong heights for the required task, may promote fatigue. Psychosocial factors, such as excessive workload and poor interpersonal relations, may cause headaches, irritability, and other symptoms, mistakenly attributed to poor IAQ. All these symptoms are similar to the ones associated with IAQ problems, further complicating the health effects assignments, blaming incorrectly the effects of the contaminated air. In reality, the physical, ergonomic, and psychosocial stressors can in fact increase dramatically the sensitivity of human beings to IAQ, generating a distorted sensation concerning the IAQ and a lower acceptability of reasonable IAQ. Therefore, these facts should be considered in the overall IAQ problem [7, 9].

3. Monitoring methodology and remediation of IAQ

Finding solutions to IAQ problems requires a methodical approach. A principle procedure for an efficient and practical investigation is presented in **Figure 1**.

Particularly, IAQ investigations for existing building could be required for two reasons [7, 8, 12]:

1. A building assessment is requested to evaluate the potential for IAQ issues. This action is generally connected to the purchase or lease of a building, when the buyer or the leaser requires an audit of the property. This may also occur as a good general practice by building owners or managers who apply the prudence in preventing problems. Most commonly, a building assessment is requested as a response to the specific complaints of the occupants.

2. The design of a new structure or renovation should be reviewed to avoid the IAQ issues with occupancy.

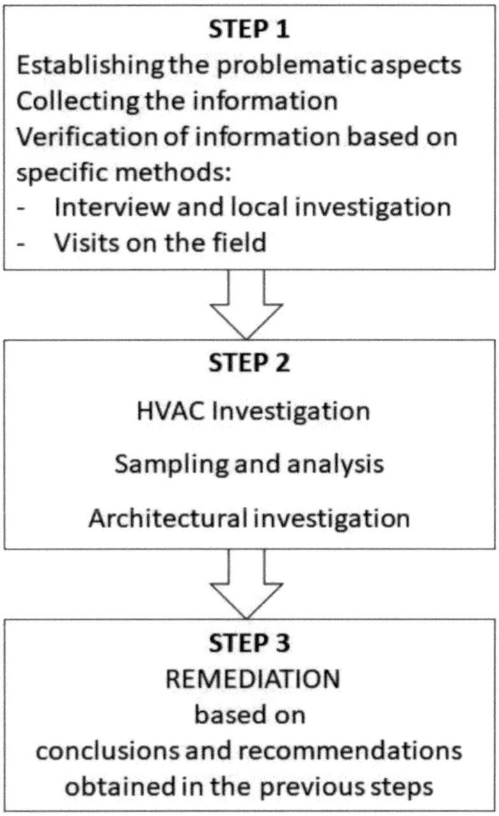

Figure 1.
Steps of monitoring and remediation of IAQ.

The building assessment is an investigative process, using three checklists [7, 12]:

1. A core checklist, with general information about the facility

2. An HVAC checklist, with information regarding the mechanical systems

3. An architectural component checklist, with information on architecture, construction, and operations

The information obtained from these checklists should allow the building investigator to reach certain conclusions about deficiencies or potential problems in the facility. These findings, along with recommendations, will make up the final report, and based on this, the remediation should be performed.

In the perspective of the remediation of a contaminated environment as described in **Figure 1**, four types of activities could be distinguished: anticipation, recognition, evaluation, and control of hazards.

Anticipating health safety problems before they really happen and acting to prevent them are the two actions which should be taken into consideration with priority by the practitioners. Foreseeing the hazardous events is possible based on previous data on the same topics, gathered from similar situations and, sometimes, statistically processed; this information could be considered historical data. The Occupational Safety and Health Administration (OSHA) considers that the so-called historical data could be used, to a certain extent and in well-defined conditions, to predict the medical problems for population [7].

The terms hazard and risk, which are used in some situations as synonyms, have in this field quit different significance. A hazard is an object or a set of conditions which can affect the human health or safety and could be extended to plants and animals in the broader context of the environment. The risk is the manifestation of an imminent harmful occurrence, defined quantitatively by a statistical expression or qualitatively via the probability of occurrence. For example, the possibility of harm from many carcinogen hazard facts is low if the exposure to the hazardous substance happens at low intensity and short duration (i.e., the risk is low). In other words, a highly toxic chemical could be out of risk if an efficient barrier is built between the substance potential emission space and the human subject working in the area. In brief, the hazard is given by the intrinsic properties of the contaminant, while the risk is rather dependent on the exposure time and intensity [7].

In this light, IAQ is an important part of so-named smart building concept, along with other essential components, such as aesthetical architecture, durability, energy savings, etc. The accomplishment of the performance objectives related to these criteria by using environment-friendly and cost-effective tools contributes to the eco-efficient built environment [13]. More details about the different strategies (e.g., from marketing to technologies, algorithms, and big data) involved in the achievement of a smart eco-efficient built environment are described in [14].

Based on these aspects, the EPA in the USA published on its website a special program, Indoor airPLUS, in order to help the new home builders to improve the quality of indoor air, by defining construction practices and product specifications that minimize the exposure to airborne pollutants and contaminants [11].

4. Main types of pollutants in indoor air living spaces and involved interactions

The international legislation and local regulations aim to protect the inhabitants and workers from harmful exposures to hazardous substances, as well as the environment from indoor-generated pollutants.

A classification of the factors causing people discomfort or morbidity could be made by the nature of these factors like chemical, physical, and biological. The route by which the harmful agent enters the body is the key solution to protect people against their harmful action. The chemical and biological contaminants enter the human body usually through inhalation, skin contact, or ingestion. Physical agents (vibration, noise, pressure, temperature, various electromagnetic radiations, etc.) usually act to the whole body and can result in harmful biochemical reactions and in tissues and organ damage [7].

The correlation between occupational exposures and worker health is studied by the environmental and occupational medicine, with specialization in industrial hygiene (an interdisciplinary science based on chemical, mechanical, civil, or environmental engineering, chemistry, biology, physics) with tight connections with air pollution; analytical chemistry; engineering; heat, pressure, ergonomics, and other physical factors; ionizing and nonionizing radiation; noise and vibration; personal protective equipment; regulations, standards, and guidelines; sampling and instrumentation; and toxicology [7].

The main types of indoor pollutants are screened below, and methods for their diction were previously published by the authors in "Electrochemical Sensors for Monitoring of Indoor and Outdoor Air Pollution" [15].

5. Air contaminants and their effects upon human health

The prediction of hazards and risks generated by some building materials is based on a review of the individual chemical components of the whole material, taking into consideration how these components interact between them as well as with other factors like humidity, respiratory gases, etc. For this reason, the safety and health file contains information on each chemical compound from the material composition, the chemistry of the possible reactions which could occur, etc.

The specific behavior upon living creatures and different characterization of pollutants from air determined the literature definition thereof as particulate pollutants and gaseous pollutants. Many terms of science and engineering will be used in this section, without a definition if the words are used consistently with commonly accepted meanings in these fields, and, therefore, could be easily found in professional references. If, however, a term has a unique or different meaning, this will be defined in the text.

Airborne contaminants could be classified according to their nature, following the definitions according to the *Glossary of Fundamentals of Industrial Hygiene* [7]:

5.1 Dusts

This word is used to designate solid particles generated by diverse mechanical and/or thermal procedures applied during exploitation and/or processing of ores, rocks, wood, and coal. Dusts are the results of advanced fragmentation associated with the above-mentioned types of operations. Generally, dusts are made of particles with high densities; therefore, the particles settle rather easily by gravity and, in most situations, do not flocculate.

5.2 Fume

This name is given for the small particles formed in air by the evaporation of solid materials (the typical case is the metal fume formed during welding). These particles have sizes less than 1 micron and needle-like morphology.

5.3 Smoke

The particles from smoke are generally formed during the combustion or sublimation processes, associated with the local overheating; the particles from smoke are called soot. The smoke cannot be usually avoided when solid carbon-rich fuels are burned (coal, wood, solid waste). In smoke, the dry soot particles are often associated with liquids in complex droplets.

5.4 Vapors

The term vapors refers to the existence of small amounts of certain substances in gas phase, substances which are, in pure state, solid or liquid at ambient temperature and pressure. When increasing the pressure and/or decreasing the temperature, most vapors transform into solid or liquid state. The vapors spread fast and easy in the air, due to their high diffusion potential associated with the high degree of freedom of moving molecules in the gas phase. Most vapors from the indoor air belong to the volatile organic compound (VOC) class (typical examples are various solvents with low boiling points).

5.5 Aerosols

The term aerosol refers to particles of 0.01–100 μm (smaller than in smokes or fumes), in liquid or solid phase. The aerosols remain dispersed in air for quite long time and have a high diffusion capacity to the lung alveoli, even when breathing. Some aerosols are beneficial to health, when a medicine is conditioned as aerosol.

5.6 Mists

Mists are formed sometimes when proper conditions are reached to condense a compound from vapor state to liquid state. The size of the droplets formed is small enough not to be settled as liquid, but remains suspended as small particles. The mists are differentiated from aerosols by the bigger size of the droplets in the first case. Mists are also formed by atomization and splashing in different liquid manipulation technologies.

The effects of the particles on the human health strongly depend on their size and behavior when entering the human body. The fine particles persist in the atmosphere for a few days without sedimentation, so they can be transported over long distances. They can have harmful effects on human health and environment even thousands of kilometers away from the source. The respiratory system supports the strongest attack by small particles. The particles that reach the lungs are fixed on the lung alveoli, reducing the oxygen exchange surface of the lungs. Three types of fractions can be defined [7]:

- The inhalable fraction, which includes all the particles that can enter the nostrils and mouth

- The thoracic fracture, which includes particles that can pass through the larynx and enter the tracheobronchial region during inhalation

- The respirable fraction, which includes small particles that can reach the alveolar region

The risk of particles is due to deposits that occur throughout the respiratory system, from the nose to the alveoli, because the respiratory system is like a channel that branches from the point of inhalation (the nose or mouth) to the pulmonary alveoli with the constant diameter decreasing. As the particle containing air passes through the tracheobronchial tract to the alveoli, the largest particles are progressively stored, followed by smaller particles. The particles with sizes less than 10 μm (PM 10) are deposited on the tracheobronchial tract, and those around 2.5 μm and smaller (PM 2.5) are stored in the lungs. They can readily be absorbed into the blood, causing poisoning or worsening of chronic respiratory diseases [7].

5.7 Gas

Gas is a state of matter characterized by materials with very low density and viscosity, expanding or contracting upon temperature and pressure changes. The gases easily diffused are distributed uniformly in any container. The gases are characterized by weak interactions between the molecules and a strong extent of freedom in moving of the molecules contained. Some gas species from the lower atmosphere layers can diffuse to the upper layers (troposphere), where they can occur chemical reactions influencing the overall atmosphere quality.

Among the gaseous contaminants of the indoor air, the nitrogen and sulfur oxides, hydrogen sulfide, carbon monoxide, carbon dioxide, as well as a large number of Volatile Organic Compounds (VOCs) could be mentioned.

Due to their health-associated issues, such indoor air contaminants are regulated by several organizations in different countries [7, 13].

Sulfur dioxide is a colorless gas with a pungent, irritating odor that causes tearing, retards breathing, and generates discomfort. At long exposure at low concentrations, it was found that pollution from sulfur dioxide and sulfates may cause a higher incidence of colon and breast cancers. In the upper respiratory system, SO_2 can cause lung edema or even death. The main sources of sulfur compounds in the atmosphere are the volcanic eruptions, the bacterial reactions in swamps, the marine phytoplankton reactions, burning of fossil fuels, and sulfide burning to produce SO_2 in the production of sulfuric acid. Traces of SO_2 result from the fuel burning, especially in diesel engines. In wet air and in the presence of UV radiation from the sun, sulfur dioxide transforms into sulfuric acid, which has a remarkable ability to associate the soot particles, generating a very toxic and stable smoke named smog [16].

Hydrogen sulfide results mainly from the urban wastewater treatment and from the spontaneous decomposition of swamp vegetation. This gas has a very unpleasant odor (as rotten eggs) and a quite acidic character [16].

Nitrogen oxides. The main oxides of nitrogen from the air are nitrogen monoxide (NO) and nitrogen dioxide. Nitrogen monoxide is a colorless and odorless gas, while nitrogen dioxide is a brownish-red gas with a strong, sulfurous odor. In combination with airborne particles, NO_2 forms a brown-red mist that, in the presence of sunlight, forms photochemical oxidants. Nitrogen dioxide is a highly toxic gas for both humans and animals because of its oxidizing effect; the degree of toxicity of nitrogen dioxide is four times higher than that of nitrogen monoxide. Short-term exposure to these pollutants causes respiratory difficulties, respiratory irritation, and lung dysfunctions. Long-term exposure at a low concentration may destroy the lung tissue, resulting in pulmonary emphysema, and exposure to high concentrations may be fatal [16].

Carbon monoxide is a colorless, odorless, and insipid gas of natural or anthropogenic provenience. The anthropogenic activities responsible for CO generation are the different processes of incomplete combustion of fossil fuels, in both high-capacity thermoelectric power plants and small domestic heating plants, as well as the combustion of fuels in internal combustion engines and chemical processes occurring in the production of iron and steel and in oil refining. CO is a very toxic gas for humans and animals, being lethal at concentrations of about 100 mg/m^3, by reducing the oxygen transport capacity in the blood. At relatively low concentrations, it affects the central nervous system, weakens the heart rate, decreases the volume of blood distributed in the body; reduces the visual acuity and physical capacity; generates acute fatigue, breathing difficulties, irritability, migraines, rapid breathing, lack of coordination, nausea, dizziness, and confusion; and reduces the ability to concentrate [17].

Carbon dioxide is the normal product of fuel combustion from the power plants, as well as from the respiratory processes of the living creatures. In ambient conditions, carbon dioxide is a colorless, odorless, slightly acidic, and non-flammable gas. Upon compression and temperature decrease, CO_2 transforms into a liquid or a solid (dry ice) [18].

Carbon dioxide is not what is called a toxic gas, but its concentration indoor is an indicator of a good, breathable healthy air. The normal outdoor CO_2 concentration is of 250–350 ppm, while indoor air of good quality can reach 1000 ppm. After this limit, the CO_2 levels begin to influence the health condition of humans: up to

2000 ppm, the sensation of poor air combines with sleepiness, between 2000 and 5000 ppm; the symptoms are drowsiness, headaches, loss of attention, and even nausea or increased heart rate. Over 5000 ppm, CO_2 could be considered a potential toxic gas, and the oxygen deprivation appears [19].

Particularly, space maximum allowable concentration of CO_2 is 10,000 ppm [2].

The *VOCs* are a numerous class of organic compounds, with small- or medium-sized molecules, present indoor especially from the construction and decorating materials, as paints, varnishes, and wax; hygiene; and beauty products for cleaning/disinfecting, degreasing, personal care, and hobby products. Commonly, the levels of the most usual VOCs are several times higher inside homes than outside, even for homes located in rural or in highly industrial areas. A characteristic of the VOCs indoor is the fact that the exposure to elevated concentrations can persist in the air long after the activity generating the chemicals had finished [20].

Among the most common VOCs, formaldehyde, aromatic and aliphatic hydrocarbons, carbonyl compounds, as well as chlorinated organic compounds are widely spread indoor [20].

Formaldehyde is one of the most common VOCs. It is a colorless gas with a sharp smell. The sources of formaldehyde indoor are lots of building and finishing materials: plywood, particleboard, glues, waterproof fabrics, foams, and fuel burning [21].

It should be noted that formaldehyde is one of the most toxic indoor air pollutants [13] and its high presence probability indoors due to the above-mentioned sources suggests the need of action in the direction of source-reduction measure implementation.

The main *aromatic, naphthenic, and aliphatic hydrocarbons* cited to be present in the indoor air, especially in the car interiors, are: benzene; toluene; ethylbenzene; xylenes; styrene; propyl-benzene; cumene; ethyl-methyl-benzenes; trimethyl-benzenes; naphthalene; methylcyclopentane; 1,2-dimethylcyclopentane; cyclohexane; dimethylcyclopentane cis- and trans-1,2-dimethylcyclopentane; methylcyclohexane; ethylcyclopentane; 1,4-dimethylcyclohexane; trimethylcyclo-hexanes; propylcyclohexane; methyl-propyl-cyclohexanes; hexane; 3,3-dimethyl-pentane; 2-methylhexane; 2,3-dimethylpentane; 3-methylhexane; 3-ethylpentane; heptane; dimethylhexanes; 2-methylheptane; octane; 2,5-dimethylheptane; 3-methyloctane; 2,6-dimethyloctane; 2-methylnonane; decane; undecane; dodecane; tridecane; tetradecane; and pentadecane.

Methylene chloride (dichloromethane), benzyl chloride, carbon tetrachloride, chloroform, trichloroethylene, tetrachloroethylene, p-dichlorobenzene, and 1,1,1-trichloroethane used together with polybrominated diphenyl ethers, hexabromocyclododecanes, tetrabromobisphenol A, and polybrominated biphenyls are some examples of *halogenated compounds* used in plastic materials to improve their technical performance or act as flame retardants [7, 10].

Radon is a radioactive, colorless, odorless, and tasteless gas [22] which occurs naturally as an intermediate step in the normal radioactive decay chains of thorium and uranium. These last elements are often found in the composition of construction materials (i.e., bricks, ceramic tiles, cement, etc.), and therefore the buildings made from these materials could accumulate in time (if the ventilation system is not appropriate) a huge amount of this radioactive gas with dangerous action for human health [23].

6. Conclusions

In order to better understand and control IAQ problems in living spaces, some general aspects related to chemical and particle content of air, with practical advices, specific terminology including links to standards and regulations, were presented in this chapter.

Monitoring methodology and remediation of IAQ were also reviewed.

Main types of pollutants in indoor air living spaces and their effects upon human health were presented.

In order to point out the importance of IAQ for everyone's health but especially to those that have chronic respiratory problems, a governmental program, namely, Indoor airPLUS, was launched in the USA, on the EPA website, consisting in a voluntary partnership and labeling actions that helps new home builders to improve the quality of indoor air by requiring construction practices and product specifications that minimize exposure to airborne pollutants and contaminants.

Acknowledgements

This work was partially supported by the following projects: EFECON, Eco-innovative Products and Technologies for Energy Efficiency in Construction; POC/71/1/4, Knowledge Transfer Partnership, Cod MySMIS: 105524, ID: P_40_295, Project co-financed by the European Regional Development Fund; and RDI Program for Space Technology and Advanced Research (STAR), project no. 157/2017, financed by Romanian Space Agency (ROSA).

Author details

Igor Cretescu[1]*, Dorina Nicolina Isopescu[2], Doina Lutic[3]* and Gabriela Soreanu[1]

1 Department of Environmental Engineering and Management, Faculty of Chemical Engineering and Environmental Protection, "Gheorghe Asachi" Technical University of Iasi, Iasi, Romania

2 Faculty of Civil Engineering and Building Services, "Gheorghe Asachi" Technical University of Iasi, Iasi, Romania

3 Department of Material Chemistry, Faculty of Chemistry, "Alexandru Ioan Cuza" University, Iasi, Romania

*Address all correspondence to: icre1@yahoo.co.uk and doilub@uaic.ro

References

[1] Soreanu G, Dixon M, Darlington A. Botanical biofiltration of indoor gaseous pollutants—A mini-review. Chemical Engineering Journal. 2013;**229**:585-594

[2] Honne A, Schumann-Olsen H, Kaspersen K, Mosebach H, Kampf D. Air quality monitoring for the International Space Station applicable to aircraft cabins and cockpits. Journal of Biological Physics and Chemistry. 2014;**14**:94-102

[3] Macatangay AV, Perry JL. Cabin. Air Quality on Board Mir and the International Space Station—A Comparison. p. 20. https://ntrs.nasa.gov/archive/nasa/casi.ntrs.nasa.gov/20070013700.pdf

[4] Roychoudhury S, Walsh D, Perry J. Microlith Based Sorberfor Removal of Environmental Contaminants. SAE 2004-01-2442; 2004

[5] Mac Donell M, Raymond M, Wyker D, Finster M, Chang YS, Raymond T, et al. Research and Development Highlights: Mobile Sensors and Applications for Air Pollutants. Report EPA/600/R-14/051; 2013

[6] Johnson SR, Garrard GG, Mitchell SM. Paper session I-C-multifunctional air revitalization systems. In: The Space Congress Proceedings. Paper 6; 27 April 1993. 1993

[7] Hays SM, Gobbell RV, Ganick NR. Indoor Air Quality: Solutions and Strategies. USA: McGraw-Hill, Inc; 1995

[8] Available from: https://www.epa.gov/indoor-air-quality-iaq/section-6-building-air-quality-guide-diagnosing-indoor-air-quality-problems

[9] Korhan O. Occupational Health. Rijeka, Croatia. ISBN: 978-953-51-2890-8: InTech; 2017. p. 258. DOI: 10.5772/63281

[10] Faber J, Brodzik K, Gołda-Kopek A, Łomankiewicz D, Nowak J, Świątek A. Comparison of air pollution by VOCs inside the cabins of new vehicles. Environment and Natural Resources Research. 2014;**4**(3):155. DOI: 10.5539/enrr.v4n3

[11] Available from: https://www.epa.gov/indoorairplus

[12] Available from: https://www.epa.gov/sites/production/files/201408/documents/sec_6.pdf

[13] Soreanu G. Chapter 12: Biotechnologies for improving indoor air quality. In: Pacheco-Torgal F, Rasmussen E, Granqvist CG, Ivanov V, Kaklauskas A, Makonin S, editors. Start-Up Creation: The Smart Eco-Efficient Built Environment. UK: Elsevier, Woodhead Publishing; 2016

[14] Pacheco-Torgal F, Rasmussen E, Granqvist CG, Ivanov V, Kaklauskas A, Makonin S, editors. Start-Up Creation: The Smart Eco-Efficient Built Environment. UK: Elsevier, Woodhead Publishing; 2016

[15] Cretescu I, Lutic D, Manea LR. Electrochemical sensors for monitoring of indoor and outdoor air pollution. In: Rahman M, Asiri AM, editors. Electrochemical Sensors Technology. Rijeka, Croatia. ISBN: 978-953-51-3193-9: InTech; 2017. pp. 855-640. DOI: 10.5772/intechopen.68512

[16] Available from: https://www.atsdr.cdc.gov/phs/phs.asp?id=251&tid=46

[17] Available from: https://www.environment.gov.au/protection/publications/factsheet-carbon-monoxide-co

[18] Available from: https://pubchem.ncbi.nlm.nih.gov/compound/carbon_dioxide#section=Physical-Description

[19] Available from: https://ohsonline.com/articles/2016/04/01/carbon-dioxide-detection-and-indoor-air-quality-control.aspx

[20] Available from: https://www.epa.gov/indoor-air-quality-iaq/volatile-organic-compounds-impact-indoor-air-quality

[21] Available from: https://www.healthlinkbc.ca/healthlinkbc-files/air-quality-VOCs

[22] Available from: https://www.epa.gov/radon/citizens-guide-radon-guide-protecting-yourself-and-your-family-radon

[23] Isopescu DN, Sluser BM, Noli F, Buckus R, Cretescu I. Monitoring of radioactivity level in autoclaved aerated concrete produced in Romania and risk assessment. Environmental Engineering and Management Journal. 2016;**15**(4):935-943. DOI: 10.30638/eemj.2016.101

Section 3

Acoustic and Thermal Comfort

Chapter 6

Noise Calculation Charts and Indoor Environmental Quality for Evaluating Industrial Indoor Environment and Health

Himanshu Dehra

Abstract

Noise, defined as "a sensation of unwanted intensity of a wave," is perception of a pollutant and a type of environmental stressor. An environmental stressor such as noise may have detrimental effects on various aspects of health. The unwanted intensity of a wave is a propagation of noise due to transmission of waves (viz. physical agents) such as sun, light, sound, heat, electricity, fluid, and fire. The effects of these physical agents on human health as noise-intruding elements in an industrial indoor environment are discussed. Noise characterization is discussed from indoor air quality and health perspective. The noise calculation charts are detailed for interference of noise waves based on a benchmark solution. These charts calculate positive and negative magnitudes of noise based on noise characterization of waves due to power difference of two intensities. The noise interference is calculated from newly devised noise measurement equations and their units. The grades and flag colors are notated to the noise calculation charts. Furthermore, illustrated examples of noise characterization calculations for indoor environment are presented using devised noise measurement equations. Indoor environmental quality and noise instrumentation are discussed. Adverse effects of pollutants on human health are summarized. Ventilation systems for dispersion of pollutants from industrial indoor environment are also elaborated.

Keywords: noise, ventilation, physical agents, noise chart, IEQ, health

1. Introduction

Improved building techniques and concern for energy efficiency have created airtight housing. These buildings also retain and recirculate indoor air-along with any contaminants. At the same time many common chemicals and materials in the home are now being implicated in conditions from chronic respiratory irritation to cancer.

These potentially harmful substances carried in indoor air fall into two groups: particles and gases [1].

Particles are invisible and evade the body's natural filtering mechanisms, carrying toxic substances deep into lung tissue and are absorbed into the body. Dust mites, pet dander, mold spores and pollen are common particles that cause asthma in some-but may cause chronic runny noses, watery eyes, headaches, lethargy, or snoring. Many people don't even know they have these allergies; they just endure them.

Gases in indoor air may also present health risks. Of most concern are the volatile organic compounds-or-VOCs-of which over 500 have been identified, and are dispersed from cleaning solutions, carpets, building materials, and many household items.

Lin et al. [2] anticipated that there will be behavioral changes that accompany population growth and aging and examined the relationship between home occupant behavior and indoor air quality. Vlek and Steg [3] discussed *the* necessity of multidisciplinary collaboration and desirable developments in environmental psychology for identifying problems, driving forces in human behavior and environmental sustainability. Stansfeld [4] discussed that in planning and health impact assessment environmental noise should be considered as an independent contributor to health risk. It has a separate and substantial role in ill-health separate to that of air pollution. Dehra [5–28] introduced characterization of physical agents as wanted physical agents and unwanted physical agents. The wanted physical agents are categorized as new field of acoustics [15]. Unwanted physical agents are categorized as noise intruding parameters, which are required to be removed from the environment by proper conditioning and filtering [18].

The aim of this chapter is to present practical understanding of how industrial environments and indoor air quality interact with noise so as to impact health and wellness. Noise, defined as "a sensation of unwanted intensity of a wave," is perception of a pollutant and a type of environmental stressor. An environmental stressor such as noise may have detrimental effects on various aspects of health. The unwanted intensity of a wave is a propagation of noise due to transmission of waves (viz. physical agents) such as sun, light, sound, heat, electricity, fluid and fire. The characterization of these physical agents on human health as noise intruding elements is discussed and also presented in mathematical form.

2. Indoor air quality and health

The following is the list of harmful materials in indoor environment:

Insulation: The jagged edged particles of fiberglass insulation are of even greater concern than asbestos, which can still be found in older homes. Urethane is a further concern.

Pressed wood in furniture and flooring: Manufactured with formaldehyde, which is released as a gas.

Dry cleaning: Source of toxic toluene and PERC.

Dust mites: The excreta and body parts of these organisms are a common allergen, a particular problem in bedrooms.

Lead: Found in older paints and plumbing pipes.

Aerosols in personal care products: A common source of organic gases, some of which are suspected carcinogens.

Gas stoves: Can be a source of combustion by-products.

Cleaning products: Can contain many harmful volatile compounds.

Pesticides: Extremely toxic when released into indoor air.

Fireplace smoke: Open fireplaces can produce ash dust, carcinogenic tars and combustion by-products like carbon monoxide and nitrogen oxides.

Wall coverings: Plasticized wallpapers can emit VOCs; wood paneling is often manufactured with formaldehyde.

Draperies: Treated fabrics can contain formaldehyde. Curtain folds help collect dust.

Tobacco smoke: Contains 43 known carcinogens, many of which are passed to non-smokers, especially in areas with poor air circulation.

Carpets: Outgases formaldehyde, harbors dust, dust mites, animal dander, and other allergens.

Paints and solvents: Sources of many volatile chemicals that are easily vaporized and absorbed by the body.

Pet dander: Animal hair and skin flakes are a common allergen.

Auto exhaust: Such fumes enter the home from the garage or nearby traffic. Carbon monoxide is the primary danger and has a cumulative effect on the nervous system.

Fuels: Storage of gasoline, kerosene, and other fuels can release volatile chemicals into the air.

VOCs and indoor air quality: VOCs are volatile organic compounds. VOCs can be released by human source in indoor environment and can react with the atmosphere to form ground-level ozone, and to a lesser degree, acid rain. Some VOCs, like the fumes from numerous interior products (glues, paints, cabinets, carpets, and pads, furniture, etc.) are toxic, and can cause a range of health problems from occasional headaches to allergic reactions, depending on the concentration and sensitivity of the individual. "Least toxic" products are those which contain levels of VOCs below the permissible levels. In some cases, a "least toxic" product may be preferable to a "non-toxic" product for reasonable performance. For additional assurance, mechanical ventilation with an air-to-air heat exchanger or outdoor duct exposed to solar radiation can be considered (examples of which are provided later) that gives fresh air without wasting heat.

Volatile organic compounds (VOCs) are found in everything from paints and coatings to underarm deodorant and cleaning fluids. VOCs have been found to be a major contributing factor to ozone, a common air pollutant which has been proven to be a public health hazard. While ozone in the upper atmosphere is beneficial, ozone at ground level is quite the opposite. The atmospheric ozone layer helps protect us from the sun's dangerous ultraviolet rays. Ground level ozone, however, is a highly reactive gas which affects the normal function of the lung in many healthy humans. These studies show that breathing air with ozone concentrations above air quality standards aggravates symptoms of people with pulmonary diseases and seems to increase rates of asthma attacks. There is also evidence that prolonged exposure to ozone causes permanent damage to lung tissue and interferes with the functioning of the immune system.

3. Noise characterization

A unified theory for stresses and oscillations is proposed by the author [7]. The following standard measurement equations are derived and adopted from the standard definitions for sources of noise interference as developed by the author [5–15].

Noise of sol: For a pack of solar energy wave, the multiplication of solar power storage and the velocity of light gives solar power intensity I. On taking logarithm of two intensities of solar power, I_1 and I_2, provides intensity difference. It is mathematically expressed as:

$$Sol = \log(I_1)(I_2)^{-1} \quad (1)$$

whereas logarithmic unit ratio for noise of sol is expressed as *Sol*. The oncisol (*oS*) is more convenient for solar power systems. The mathematical expression by the following equality gives an oncisol (*oS*), which is 1/11th unit of a *Sol*:

$$oS = \pm 11 \log(I_1)(I_2)^{-1} \quad (2)$$

Noise of therm: For a pack of heat energy wave, the multiplication of total power storage and the velocity of light gives heat power intensity *I*. The pack of solar energy wave and heat energy wave (for same intensity *I*), have same energy areas, therefore their units of noise are same as *Sol*.

Noise of photons: For a pack of light energy beam, the multiplication of total power storage and the velocity of light gives light power intensity *I*. The pack of solar energy wave and light energy beam (for same intensity *I*), have same energy areas, therefore their units of noise are same as *Sol*.

Noise of electrons: For a pack of electricity wave, the multiplication of total electrical storage and the velocity of light gives electrical power intensity *I*. The pack of solar energy wave and electricity wave (for same intensity *I*), have same energy areas, therefore their units of noise are same as *Sol*.

Noise of scattering: For a pack of fluid energy wave, the multiplication of total power storage and the velocity of fluid gives fluid power intensity *I*. On taking logarithm of two intensities of fluid power, I_1 and I_2, provides intensity difference. It is mathematically expressed as:

$$Sip = \log(I_1)(I_2)^{-1} \quad (3)$$

whereas, logarithmic unit ratio for noise of scattering is *Sip*. The oncisip (*oS*) is more convenient for fluid power systems.

The mathematical expression by the following equality gives an oncisip (*oS*), which is 1/11th unit of a *Sip*:

$$oS = \pm 11 \log(I_1)(I_2)^{-1} \quad (4)$$

For energy area determination for a fluid wave, the water with a specific gravity of 1.0 is the standard fluid considered with power of $\pm 1\ Wm^{-2}$ for a reference intensity I_2.

Noise of scattering and lightning: For a pack of fire wave, the intensity, *I*, of fire flash with power of light is the multiplication of total power storage and the velocity of light. Whereas for a pack of fire wave, the intensity, *I*, of fire flash with power of fluid, is the multiplication of total power storage capacity and velocity of fluid.

- For a noise due to fire flash, the collective effect of scattering and lightning is obtained by superimposition principle.

- For same intensity *I*, the pack of solar energy wave and a fire flash with light power have same energy areas; therefore, their units of noise are same as *Sol*. The therm power may also be included in fire flash with power of light.

- For same intensity *I*, the pack of fluid energy wave and a fire flash with fluid power have same energy areas; therefore, their units of noise are same as *Sip*.

In determining the areas of energy for the case of fluids other than water, a multiplication factor in specific gravity has to be evaluated.

Noise of elasticity: For a pack of sound energy wave, the product of total power storage and the velocity of sound gives sound power intensity I. On taking logarithm of two intensities of sound power, I_1 and I_2, provides intensity difference. It is mathematically expressed as:

$$Bel = \log (I_1)(I_2)^{-1} \tag{5}$$

whereas, logarithmic unit ratio for noise of elasticity is *Bel*. The oncibel (oB) is more convenient for sound power systems. The mathematical expression by the following equality gives an oncibel (oB), which is 1/11th unit of a *Bel*:

$$oB = \pm 11 \log (I_1)(I_2)^{-1} \tag{6}$$

There are following elaborative points on choosing an *onci* as 1/11th unit of noise [15]:

i. Reference value used for I_2 is $-1\ \text{Wm}^{-2}$ on positive scale of noise and $1\ \text{Wm}^{-2}$ on negative scale of noise. In a power cycle, all types of wave form one positive power cycle and one negative power cycle [9]. Positive scale of noise has 10 positive units and 1 negative unit. Whereas, negative scale of noise has 1 positive unit and 10 negative units;

ii. Each unit of sol, sip, and bel is divided into 11 parts; 1 part is 1/11th unit of noise.

iii. The base of logarithm used in noise measurement equations is 11.

iv. Reference value of I_2 is $-1\ \text{Wm}^{-2}$ with I_1 on positive scale of noise and should be taken with negative noise measurement expression (see Eqs. (2), (4) and (6)); therefore, it gives positive values of noise.

v. Reference value of I_2 is $1\ \text{Wm}^{-2}$ with I_1 on negative scale of noise and should be taken with positive noise measurement expression (see Eqs. (2), (4) and (6)); therefore, it gives negative values of noise.

3.1 Estimating changes in noise power and noise pressure levels

In some cases, it is difficult to measure intensity of a power source; therefore pressure p can be measured. The relationship between pressure and intensity is given by:

$$I = \frac{p^2 rms}{\rho c} \tag{7}$$

where, p = root mean-square (rms) pressure, N/m^2, ρ = density of wave medium, kg/m^3, c = speed of wave, m/s.

Equations (2), (4), (6) are re-written in the form:

$$\text{Intensity difference } \Delta I = \pm 22 \log \frac{p}{po} \tag{8}$$

The addition of two equal pressures results in an increase of:

$22 \log 112 = 6.4$ onci sol (oS) and addition of two equal powers result in an increase of 3.2 onci sol. When two equal pressure levels are added, we are adding in effect two equal power levels, therefore:

$$Lp1 + Lp2 = 11 \log_{11} \left(\frac{p}{pref} \right)^2 + 3.2oS \tag{9}$$

Similarly, it can be shown that when N identical noise sources are added,

$$Lp(total) = Lp1 + 11 \log_{11} N \tag{10}$$

11 log N is plotted as a function of N in **Figure 1**.

Table 1 shows how to add two unequal noise levels and **Figure 2** presents **Table 1** graphically.

Table 2 has also notated grades and flag colors under limiting conditions [15]. **Figure 3** has presented a double-sided hexagonal slide rule with seven edges for noise measurement representing seven sources of noise. Reference value used for I_2 is -1 Wm^{-2} on positive scale of noise and 1 Wm^{-2} on negative scale of noise. Positive scale of noise has 10 positive units and 1 negative unit, whereas negative scale of noise has 1 positive unit and 10 negative units. Each unit of sol, sip and bel is divided into 11 parts, 1 part is 1/11th unit of noise. The base of logarithm used in noise measurement equations is 11.

The results of noise filtering using various noise measurement equations for an outdoor duct exposed to solar radiation are tabulated in **Tables 3–7**. **Table 8** has presented noise calculation charts based on intensity and pressure differences so as to calculate onci sol, onci sip and onci bel.

3.2 Thermal environment

Temperature of the ambient air has a great influence on the occupant's physical state and his work efficiency. Air conditioning can prevent excessive cold and heat because high temperature in combination with high degree of humidity cause

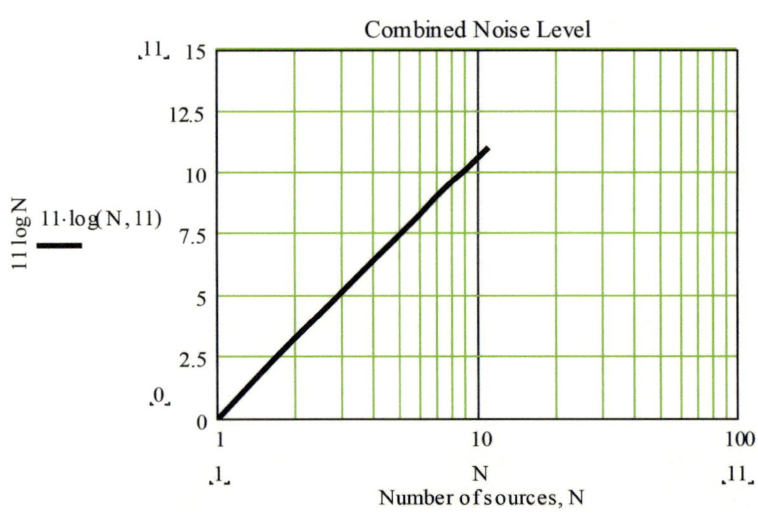

Figure 1.
Predicting the combined noise level of identical sources.

Difference between two levels, oncisol, oncisip, oncibel	Add to the higher level, oncisol, oncisip, oncibel
0	3.18
1	1.86
2	1.319
3	1.024
4	0.836
5	0.708
6	0.612
7	0.54
8	0.484
9	0.437
10	0.399
11 or more	0

Table 1.
Addition of unequal noise levels.

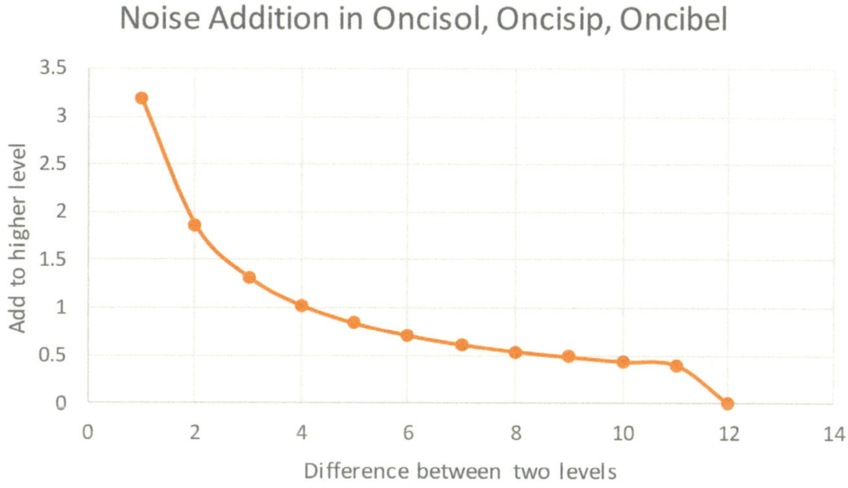

Figure 2.
Noise addition.

premature fatigue, overheating and excessive sweating. Interestingly, air temperature appears to increase 1–2°C and even more with every subsequent meter above the floor level and may reach 40–50°C near the ceiling. Low temperatures are likely to cause local or general cooling of the human body and lead to catarrhal and other respiratory disorders. Humidity is an important environmental characteristic which determines the optimum safe temperature in the work zone, other being the physical effect that the work may demand. The air remains humidified, no matter how dry it feels, by water vapors it invariably contains. The concept of air humidity differentiates: maximum humidity which is the uppermost quantity of moisture that may be held in the air at a given temperature; absolute humidity, that is the actual quantity of moisture held in the air at a given temperature; and relative humidity, defined as the percentage ratio of the absolute humidity to the maximum humidity at a given temperature. The hygienic and sanitary standards refer always

Grades	Noise grades and flag colors under limiting conditions		
	Noise of sol	Noise of scattering	Noise of elasticity
G_2[a] = ±U	Sol	Sip	Bel
$G_1 = G_2 = U$	No positive solar energy	No positive fluid energy	No positive sound energy
Base color for $G_1 = G_2$			
$G_1 = U$ $+ \rightarrow 0\ Wm^{-2}$	Decreasing solar energy	Decreasing fluid energy	Decreasing sound energy
Base color for G_2			
$G_1 = +ve$	Increasing solar energy	Increasing fluid energy	Increasing sound energy
Base color for G_2			
$G_1 = -U\ Wm^{-2}$	Negative solar energy	Negative fluid energy	Negative sound energy
	Darkness	Low pressure	Inaudible range
Base color for G_2			
$G_1 = -ve$	Darkness increasing, distance from point source of light increasing	Low pressure increasing, vacuum approaching	Inaudible range increasing, vacuum approaching
Base color for G_2			
$G_1 = -U$ $+ \rightarrow 0\ Wm^{-2}$	Negative solar energy	Negative fluid energy	Negative sound energy
	Decreasing darkness	Decreasing low pressure	Decreasing inaudible range
Base color for G_2			

[a]*Reference value of $G_2 = \pm U$ signifies the limiting condition with areas of noise interference approaching to zero.*

Table 2.
Noise grades and flag colors under limiting conditions.

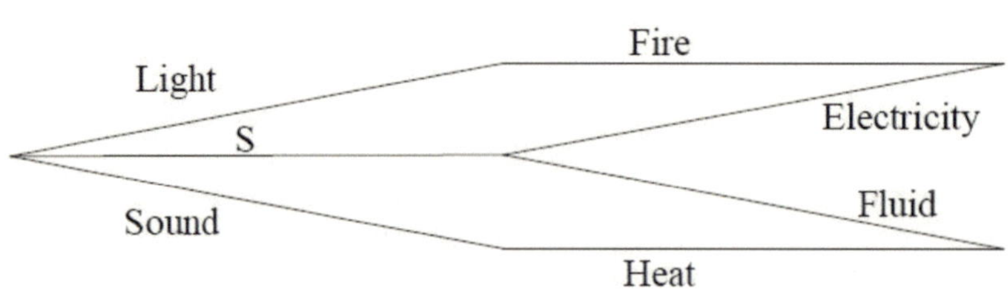

Figure 3.
A double sided hexagonal scales of noise with seven edges (S denotes sun).

to relative humidity as a factor conducive to the safe working environment. Excess humidity may result from various processes that produce water vapors into the working atmosphere. It is generally accepted that the optimum relative humidity is 40–60%. Excessive saturation of the air with water vapor prevents vaporization of moisture from the skin and the lungs which causes much discomfort and reduces work efficiency. Vaporization of excess moisture that results from hyperhidrosis plays an important part in heat transfer from the human body. Vaporization or

Property	Value	Property	Value
Solar irradiation	650 Wm^{-2}	Width of air gap	0.025 m
Ambient heat transfer coefficient	13.5 $Wm^{-2}\,K^{-1}$	Thermal conductivity of air	0.02624 $Wm^{-1}\,K^{-1}$
Ambient air temperature	−5°C	Specific heat of air (cp)	1000 J $kg^{-1}\,K^{-1}$
Building space temperature	20°C	Density of air	1.1174 kg m^{-3}
Height of duct	3.0 m	Kinematic viscosity of air	$15.69 \times 10^{-6}\,m^2\,s^{-1}$
Width of duct	1.0 m	Prandtl number of air	0.708
Thickness of outer wall of duct	0.0025 m	Air velocity for obtaining mass flow rate	0.75 m s^{-1}
Absorbance of outer wall with flat black paint	0.95	Stefan Boltzmann constant for surface of duct walls	5.67 × 10–8 $Wm^{-2}\,K^{-4}$
Thermal conductivity of aluminum alloy for HVAC duct	137 $Wm^{-1}\,K^{-1}$	Emissivity of back surface of duct walls	0.95
RSI value	1.0 $m^2\,K\,W^{-1}$	Number of nodes in x-direction	$Nx = 3$
Thickness	0.04 m	Number of nodes in y-direction	$Ny = 10$, $\Delta y = 0.3$ m

Table 3.
Properties of physical domain.

Solar irradiation (Wm^{-2})	Air temperature difference (ΔT) °C	Noise of sol oS (oncisol)
450	15.50	28
550	18.90	28.93
650	22.40	29.7
750	25.90	30.36
850	29.40	30.91

Table 4.
Temperature difference and noise of sol with solar irradiation (air velocity: 0.75 ms^{-1}).

Air velocity (ms^{-1})	Fluid power (Wm^{-2})	Air temperature difference (ΔT) °C	Noise of scattering *oS* (oncisip)
1.35	47.62	15.28	17.72
1.05	37.0	18.22	16.50
0.75	26.45	22.40	15.02
0.45	15.87	28.15	12.65
0.15	05.29	29.80	07.64

Table 5.
Temperature difference and noise of scattering with air velocity (S = 650 Wm^{-2}).

evaporative heat transfer brings relief and prevents unnecessary fatigue and discomfort. Likewise, a decreased level of humidity, e.g., up to 20% also brings much discomfort to the occupant: he feels very dry in the upper respiratory tract and the mucosal membranes. Movement of the air about the work zone is an important factor of a comfortable working environment. It will be noted that the human body

(ΔT) °C	Mass flow rate ($kg\ s^{-1}$)	Thermal power (Wm^{-2})	Noise of therm *oS* (oncisol)	(ΔT) °C	Mass flow rate ($kg\ s^{-1}$)	Thermal power (Wm^{-2})	Noise of therm *oS* (oncisol)
15.50	0.01376	71.09	19.5602	15.28	0.0231	117.65	21.868
18.90	0.01275	80.325	20.119	18.22	0.0171	103.85	21.296
22.40	0.0120	89.6	20.614	22.40	0.0120	89.6	20.614
25.90	0.0115	99.2833	21.043	28.15	8.1×10^{-3}	76.0	19.866
29.40	0.0111	108.78	21.505	29.80	6.2×10^{-3}	61.59	18.898

Table 6.
Mass flow rate and noise of therm with (ΔT) °C.

Air velocity (ms^{-1})	Fluid power (Wm^{-2})	Noise of scattering *oS* (oncisip)	Sound pressure (Nm^{-2})	Sound power intensity (Wm^{-2})	Noise of elasticity *oB* (oncibel)
1.35	47.62	17.72	557.5	752.7	30.36
1.05	37.0	16.50	433.65	455.33	28.05
0.75	26.45	15.02	309.75	232.31	24.97
0.45	15.87	12.65	185.85	83.63	20.24
0.15	05.29	07.64	61.94	09.29	10.12

Table 7.
Noise of elasticity with air particle velocity (impedance Z_o = 413 N s m^{-3} at 20°C).

feels slight movements of the air at flow velocities about 0.15 m/s, and at temperatures up to 36°C the effect of movement is refreshing, while at 40°C and higher, it is suppressing. It is commonly believed that the optimum comfortable temperature in the indoor environment in a temperate climate is 20°C. It will be however more correct to say that the optimum safe temperature depends on a number of factors including relative humidity, physical effort, air movement, etc. All the environmental factors are closely interrelated.

3.3 Noise instrumentation for sensing physical agents of indoor environment

Noise instrumentation can be classified as per type of physical agent (light, sound, heat, fluid, electricity, fire and sun) in indoor environment for which monitoring is required. To ensure proper environmental control, the climatic parameters within an enclosure should be periodically checked and measured. Various kinds of measuring instruments can be used for this purpose.

Thermometers, placed on the walls or columns 1.5 m high above floor level and not closure than 1 m to the heating device, are used to measure the air temperature. Careful temperature control is necessary. In such cases, thermographs are used for continuous recording of the air temperature in the workroom. Psychrometers or wet and dry hydrometers, are devices used to determine the relative humidity of the air by the difference in the readings of a pair of similar thermometers mounted side by side. The one is dry bulb, and the other has its bulb wrapped in a damp wick dipping into distilled water. The rate of evaporation of water from the wick and the consequent cooling of the wet bulb is dependent on the relative humidity of the air which can be obtained by using a psychrometric table, or a nomogram from readings of the two thermometers. Use is also made of stationary aspiration

a	b	Intensity ratio (11^a)	Pressure ratio (11^b)	←oSol→ ← oSip → ← oBel→	Pressure ratio $(1/11)^b$	Intensity ratio $(1/11)^a$
0	0	1	1	0	1	1
1/11	1/22	1.244	1.115	±01	0.897	0.804
2/11	2/22	1.546	1.244	±02	0.804	0.647
4/11	4/22	2.392	1.546	±04	0.647	0.418
6/11	6/22	3.699	1.923	±06	0.520	0.270
8/11	8/22	5.720	2.392	±08	0.418	0.175
10/11	10/22	8.845	2.974	±10	0.336	0.113
12/11	12/22	13.679	3.699	±12	0.270	0.073
14/11	14/22	21.155	4.599	±14	0.217	0.047
16/11	16/22	32.715	5.720	±16	0.175	0.031
18/11	18/22	50.594	7.113	±18	0.141	0.020
20/11	20/22	78.242	8.845	±20	0.113	0.013
22/11	22/22	121.000	11.000	±22	0.091	8.264×10^{-3}
24/11	24/22	187.124	13.679	±24	0.073	5.344×10^{-3}
26/11	26/22	289.383	17.011	±26	0.059	3.456×10^{-3}
28/11	28/22	447.525	21.155	±28	0.047	2.235×10^{-3}
30/11	30/22	692.089	26.308	±30	0.038	1.445×10^{-3}
32/11	32/22	1070	32.715	±32	0.031	9.343×10^{-4}
34/11	34/22	1655	40.684	±34	0.025	6.042×10^{-4}
36/11	36/22	2560	50.594	±36	0.020	3.907×10^{-4}
38/11	38/22	3959	62.917	±38	0.016	2.526×10^{-4}
40/11	40/22	6122	78.242	±40	0.013	1.633×10^{-4}
42/11	42/22	9467	97.300	±42	0.010	1.056×10^{-4}
44/11	44/22	14,640	121.0	±44	8.264×10^{-3}	6.830×10^{-5}
46/11	46/22	22,640	150.47	±46	6.646×10^{-3}	4.417×10^{-5}
48/11	48/22	35,020	187.12	±48	5.344×10^{-3}	2.856×10^{-5}

a	b	Intensity ratio (11^a)	Pressure ratio (11^b)	←oSol→ ← oSip → ← oBel→	Pressure ratio $(1/11)^b$	Intensity ratio $(1/11)^a$
50/ 11	50/ 22	54,150	232.70	±50	4.297×10^{-3}	1.847×10^{-5}
66/ 11	66/ 22	1.772×10^6	1331	±66	7.513×10^{-4}	5.645×10^{-7}
77/ 11	77/ 22	1.949×10^7	4414	±77	2.265×10^{-4}	5.132×10^{-8}
88/ 11	88/ 22	2.144×10^8	14,640	±88	6.830×10^{-5}	4.665×10^{-9}
99/ 11	99/ 22	2.358×10^9	48,560	±99	2.059×10^{-5}	4.241×10^{-10}
110/ 11	110/ 22	2.594×10^{10}	161,100	±110	6.209×10^{-6}	3.855×10^{-11}

Example: to find oSol corresponding to a pressure ratio of 363
Ratio of 363 = 11 × 33
In oSol = +22 + 32 oSol
= +54 oSol

Table 8.
Noise calculation chart estimating onci sol, onci sip, and onci bel.

psychrometers which are equipped with a fan to draw the air through the device and increase the accuracy of measurement.

Hydrographs are used for continuous recording of air humidity where the humidity requirements are most stringent. Anemometers are instruments for measuring the velocity of the air. The common types are revolving-vane and revolving-cup anemometers. A common anemometer of the vane type consists of eight vanes fixed on a bulb at 45° to the air stream and pivoted so as to be capable of rotation in a vertical plane. The speed of rotation is indicated on a dial calibrated to read air velocity from 0.3 to 5 m/s. The cup anemometer differs from the vane type anemometer in that it consists of four semi-spherical cups carried on the ends of four radial arms pivoted soas to be capable of rotation in a horizontal plane. The speed of rotation is indicated on a dial graduated to give velocities of air stream from 1 to 20m/s. Airstream velocities under 0.3 m/s are measured by means of a micro-anemometer or electrical-thermal anemometer. Pressure tubes are used to measure both pressure (total and static) and velocity of the air in air ducts. The dynamic (velocity) pressure is determined as the difference between the total and the static pressures. The air velocity in air ducts can be measured with a pressure head device (which is combination of a static and pitot tube connected to the opposite sides of a differential pressure U-gage to give a visual reading corresponding to the speed of an air current).

Indication tube is used for measuring contents of air contaminants (largely toxic vapors and gases, such as, CO, SO_2, nitrogen oxides, ethanol, etc.). A common type is hermetically sealed glass tube about 4–7 mm wide and 100 mm long containing a filler (crushed silica-gel, glass, or porcelain crumps) treated with solutions of variable reagents. It also has chemical absorbents to bind unsuitable analysis, a measured quantity of air is brought into intimate contact with the contents of the open tube for a certain time interval, and the concentrations of the controllable impurity can be read on a scale by length or rate of change in color of the filler material that has completed reaction.

Purity of the air can also be measured by air or gas analyzers of certain design. Direct techniques of gas analysis-spectrometry, electrical-chemical, and optical methods, permit the analysis of the air to be performed continuously and automatically. The sampling method implies that samples of air are analyzed in a laboratory by trained specialists. Although the process is lengthy and costly it provides accurate measurements necessary for the adequate air control.

Dust contents in the air of indoor environment are determined by passing a measured quantity of air through filters during a particular time interval and calculating the mass of dust thus collected. There exist also several other proximate methods of quick determination of the character of dust and size of dust particles.

3.4 Checking human noise behavior

Noise behavior is checked by identifying a source and a sink of noise, i.e., a person making noise in the environment and a person affected by such noise in the environment. Behavior of human beings is controlled by the central nervous system. Neurology is the study of nervous system. The behavior of human beings is studied in psychology. Psychophysics is a study in which physical stimuli are perceived by human beings. Psychophysiology is the field of study of the interaction between environmental stimuli and physiological functions of the body. The study of abnormal behavior is dealt by a psychiatrist. Abnormal human noise behavior interferes in proper functioning and wellness of the individual and society.

In order to monitor noise and human noise behavior in indoor environment, these noise characterization techniques mentioned in previous sections can be used. For example, a monitoring of an improved built environment for sensing integrated noise parameters would not only result in energy conservation, economical affordability but also result in generating less noise pollution of indoor pollutants mentioned earlier. Sensors and transducers for measurement, monitoring and control of noises (light, sound, heat, fluid, electricity, fire and sun) can be electromagnetic, thermoelectric and piezoelectric, etc. With integrated environmental control, these environmental parameters can be characterized and checked for comfort and wellness and controlled through various environmental monitoring sensors. The effect of these human behavioral parameters is characterized on a logarithmic noise scale.

Biomedical instrumentation for measurement and monitoring of human noise behavior from human systems of indoor environment require real time informatics capabilities. Sensing and actuating capabilities as well as measurement systems for noise can be "in vivo" and "in vitro." The signals can be classified as bio-electric, bio-sound sample, bio-mechanical, bio-chemical, bio-magnetic, bio-optic and bio-impedance depending upon origin of "stresses and oscillations" in a noise system.

Use of biological sensors: all living organisms contain biological sensors with functions similar to those of the electro and mechanical sensors. Most of these are specialized cells that are sensitive to: light, motion, temperature, magnetic fields, gravity, humidity, vibration, pressure, electrical fields, sound and other physical aspects of the external environment and also physical aspects of the internal environment such as stretch motion of the organism and position of appendages (proprioception) an enormous array of environmental molecules including toxins, nutrients and pheromones and many aspects of the internal metabolic milieu, such as, glucose level, oxygen level or osmolality, an equally varied range of internal signal molecules such as hormones, neurotransmitters and cytokines and even the differences between proteins of the organism itself and the environment or alien creatures. The human senses are examples of specialized neuronal sensors.

3.4.1 Psychophysiological measurements

With proper psychophysiological measurements it is possible to detect source and sink of noise, i.e., a person making noise in the environment and a person affected by such noise in the environment. One of the ways to study noise behavior is to measure electrical signals of the brain and nervous system. The voltages recorded in EEG are the result of many processes that occur in the brain. It is very difficult to obtain signals related to particular function. Autonomic nervous system controls many body functions like blood pressure, heart rate, perspiration and salivation. The autonomic nervous system cannot be controlled, but will be influenced by external stimuli and emotions of a person. On recording these body functions one can get insight into emotional changes that cannot be measured directly but can be estimated from these recordings. One of the practical applications of this principle is the polygraph. It is a device for recording several body functions simultaneously that may show changes when interrogation causes anxiety in the person. It is the recording of pulse, blood pressure, respiratory rate, sweating (skin resistance), EEG, heart rate, plethysmographs, temperature, EEG, EMG and EOG simultaneously through multiple channels. A polygraph is otherwise called a lie detector as it is sometimes used to detect if a person is lying. The principle behind this is that the fright causes sweating which may reduce skin resistance response, increase heart rate and increase blood pressure. The judicial system has not yet accepted the polygraph as a foolproof system.

Testing motor responses: Response of the skeletal muscles or simply motor responses are under voluntary control. But it requires learning phase for the proper functioning with coordination of adjacent muscles. Many devices are available in the literature and in the market to measure motor responses and to study the influence of factors like fatigue, stress and/or for drugs taken. The devices used to test are simple as follows: manual dexterity test instrument consists of small objects to be assembled by the patient. The time required to assemble the objects is the measure of motor response. Another instrument, called steadiness testers consist of a metal stylus, has to be moved through narrow channels of different shapes, without touching the adjacent walls. For every contact of stylus with wall increment a counter. The counter reading is a measure of motor response. The third method of measuring motor response is that a light spot moves with adjustable speed and adjustable shape pattern (circular/rectangular/star shaped). The patient has to track the spot with a photosensitive probe. An indicator and timer measure the percentage of duration on which the patient is on target during a certain test interval. Similarly, the performance of certain muscles or muscle groups can be monitored and measures with the help of dynamometers which measure the force exerted by the patient.

Sensory measurements: Human sensors provide information inputs required by the person to orient him-self in the desired manner. There are many methods and instruments available to measure the performance of the sense organs. No sophisticated equipment is required for some of these sensory measurements. Ability of sensing of temperature can be measured by having different objects at different temperatures or water baths at different temperatures. Touch perception can be measured by stimulating the skin with bristles of horse hair. Complex devices are required to measure optical perception. For example, ability of a person to view a spot with respect to variable brightness and size of the spot as well as background brightness. A special device to study visual perception is "tachistoscope." In this case, two objects are shown alternatively with adjustable time interval. This can be achieved by illuminating the objects one at a time or by mechanical shutters. The ability of the person to recognize the objects and time duration for switching between the objects are measure of perception of eyes.

Experimental analysis of the behavior: Numerical data is required to describe and analyze behavior accurately. Especially for a mathematical analysis numerical data is required. Many basic behavioral experiments are conducted with animals like rats, pigeons and monkey as subjects. These basic behavioral experiments are conducted in skinner boxes (soundproof enclosure) in which animals are isolated from the external environment. Behavior expressed by a subject to interact with and to modify the environment is called instrumental or operant behavior. Instrument behavior that is positively reinforced tends to occur more frequently in future, whereas behavior that is negatively reinforced decreases in frequency. In the case of animal experimental study, positive reinforcement is usually done in the form of food given to animals. Negative reinforcement is in the form of harmless, but painful electric shocks through the floor of the cage. On selecting suitable reinforcement, the behavior of the animal can be studied. In this case, the skimmer box should be equipped with response bar and stimulus light. A response of the animal for a given stimuli will give pressure on the response bar. The pressure on the bar will be sensed electronically and can be stored or can be plotted on a recorder. The results obtained on animal experiments can be extrapolated on human behavior. Some of the human behaviors can be explained as having been reinforced by the social environment.

Biofeedback instrumentation: Feedback signal is used to control a process. The output of a system is measured and compared with reference signal, based on the difference; necessary action is taken to bring the output of the system very close to reference value. These types of systems are called feedback systems. If this concept is applied to biological processes within body, it is called biological feedback or biofeedback. Many physiological processes have been evaluated for possible control through biofeedback methods. The physiological processes that can be controlled by biofeedback methods are EEG, EMG, heart rate and blood pressure. It has also been observed that the alpha wave is more prevalent when a person is mediating or being with closed eyes. Heart rate and blood pressure can be controlled to a certain degree by biofeedback methods. Biofeedback instrumentation includes a transducer and amplifier to measure the variable to be controlled. The magnitude of the measured variable is converted into suitable visual or audio signals, and then is presented to the person as feedback. Biofeedback has been defined as the purest form of self-control. The success of biofeedback depends on how the data is interpreted and also depends on the training of the person.

3.5 Indoor air quality and noise

With the increased interest in indoor air quality and the need to monitor potentially dangerous gases, gas concentration measurements have become increasing more prevalent in building DDC system design. Many gas measuring devices are currently available for use in HVAC applications [29].

There are many types of gas measuring devices available for use with DDC systems. Currently, the three most common gases measured in HVAC applications are carbon monoxide, carbon dioxide, and refrigerant gases. These gas measuring devices can be modified for measuring noise of scattering (oncisip) based on concentration and fluid power of these gases. Some examples illustrating noise of scattering (oncisip) are mentioned in previous sections.

Carbon monoxide: Carbon monoxide is a poisonous gas that is most commonly generated as the byproduct of the incomplete combustion of carbon based fuels. Carbon monoxide is generated by all fuel burning equipment including internal combustion engines. Carbon monoxide detectors are used to operate ventilation equipment to prevent carbon monoxide levels from becoming unsafe. They are also used to warn facility owners and occupants of unsafe levels in garages, loading docks, tunnels and other areas where vehicles are operated. Solid state sensing technology is most commonly used. Simple or multiple sensing point versions are available that can provide contact closures at one or more set levels and/or analog signals that are proportion to carbon monoxide concentration.

Carbon dioxide: Carbon dioxide is a non-toxic gas produced by the respiration of living organisms, by the complete combustion of carbon, and by photosynthesis in green plants. Carbon dioxide exists in the air in the amount of 320–350 ppm. Carbon dioxide concentration inside buildings has been related to general ventilation adequacy and is commonly monitored by DDC control systems as a measure of indoor air quality and ventilation adequacy. It is also measured by building DDC systems and used to control outdoor air fans and dampers to keep concentration below set levels. The most commonly used sensing technology is Non-Dispersive-Infra-Red (NDIR). This is based on the principle that carbon dioxide gas absorbs infrared radiation at the 4.2 μm wavelength. Attenuation of an infrared source can be related to the gas concentration in air in the range of 0–5000 ppm with a general accuracy of plus or minus 150 or 50 ppm over narrower ranges.

Refrigerant gas: Refrigerant gas detectors have been in widespread use since safety codes for mechanical refrigeration required their use in the operation of emergency ventilation systems to evacuate hazardous concentrations of refrigerant gas in machinery rooms and other applicable enclosed areas. Detectors broadly sensitive to families of CFC and HCFC gases commonly used, as refrigerants are available. Gas specific detectors are also available to detect individual refrigerant gases including CFC, HFC, HCFC and ammonia specific to the equipment in use. The most commonly used are infrared (IR), photo-acoustic, and solid state sensing technologies. Single or multiple sensing point versions are available that can provide contact closures at one or more set levels and/or analog signals that are proportional to refrigerant concentration.

4. Effects on human health

It has been proved beyond doubt that polluted air is highly detrimental to human health and is a contributory factor in deaths from diseases, such as lung cancer [1]. Bad air quality is mostly associated with respiratory diseases ranging from common cold to lung cancer. Both gaseous and particulate pollutants cause severe damage to the respiratory system leading to emphysema, bronchitis and asthma. Polluted air irritates the eyes and some pollutants like lead tend to accumulate in the body, at times to dangerous levels. Deaths from lung cancer are on the increase and the prime causative agent is suspected to be polluted air. It may be noted that the rate of lung cancer is twice in large cities in comparison to rural areas. An important point to mention here is that children are more susceptible to polluted air than grown-ups. The prominent gaseous pollutants are carbon monoxide, sulfur dioxide, oxides of nitrogen, hydrogen sulfide and certain acids, aldehydes and hydrocarbons. The particulate pollutants include dust, silicious matter and asbestos.

Carbon monoxide is an asphyxiant gas which when inhaled reacts with the hemoglobin in blood to reduce the oxygen-carrying capacity of blood. Persons already suffering from diseases like anemia are more prone to get affected because carbon monoxide may lead to serious injuries to vital organs. Sulfur dioxide is the most serious and widespread air pollutant. It has been reported that lower levels of concentration of sulfur dioxide cause temporary spasm of the smooth muscles of the bronchia. Higher concentration induces increased mucus production. The cilia which protect the respiratory system get affected by sulfur dioxide causing cough, shortness of breath and spasm of the larynx. It may also cause acute irritation of the eye membranes resulting in tears and redness.

Oxides of nitrogen are pulmonary irritants and excess concentrations of which may lead to pulmonary hemorrhage. Hydrogen sulfide is well known for its characteristic rotten-egg odor. Hydrocarbons emitted from automobile exhaust reportedly cause lung cancer. Lead, which is discharged into the air via the automobile exhaust, is a cumulative poison that may cause brain damage in children. It interferes with the development and maturation of red blood cells. An interesting point to note is that the amount of lead in blood is higher in smokers than in non-smokers.

Among the particulate pollutants asbestos needs a special mention. The use of asbestos needs a special mention. The use of asbestos has increased several folds during 1970s–1990s. Used primarily as roofing material for buildings, and for insulating brake and clutch linings in automobiles, asbestos content in the air has also increased. Accumulation of asbestos can lead to the dreaded disease, asbestosis—a severe scarring of the lungs.

Another dreaded disease caused by polluted air (containing dust particles) is silicosis. Silicosis is the most common dust-related disease. It results from inhaling quartz dust or particles of other silica-containing rocks. Workers in mines, and industries like that of pottery, ceramics, granite carving and sand basting, run a risk of contracting this disease which in early stages shows no subjective symptoms but gradually leads to cough and shortness of breath. Silicosis is often combines with tuberculosis. The disease is gradually progressive and death occurs due to heart failure or pulmonary tuberculosis. Other serious dust diseases include block-lung disease (from inhaling coal dust), berylliosis (from beryllium dust), and byssinosis (cotton dust).

A summary of pollutants and their effects on human health is given in **Table 9** for ready reference.

5. Ventilation systems

The dispersion of pollutants from working and production areas depends very much on the proper choice of a ventilation system [30]. Ventilation systems require careful designing. This applies particularly to exhaust ventilation which, if badly designed, can be worse than no ventilation at all. Exhaust hoods or slots should be so located that no part of the fumes or dusts being removed can enter the worker's breathing zone. At the same time, ventilation should ensure that other conditions (equipment, structural elements of buildings, etc.) satisfy the requirements of the manufacturing process. Ventilation in industrial buildings and other auxiliary facilities is mandatory. Ventilation is the process of replacement of vitiated air by fresh air. Depending on how the air currents are induced, ventilation can be natural or drought and mechanical. A combination of the two types is known as the compound or hybrid ventilation.

Mechanical ventilation, depending on the direction of induced currents, can be exhaust, forced (plenum), and plenum-exhaust. By scope of action, distinction must be made between the general or dilution ventilation and local ventilation. Proper choice of ventilation system is essential for ensuring continuous supplies of good quality air into the working areas. Exhaust ventilation, for example, is one means of removing dusts, vapors and corrosive gases. Determining proper places of their possible releases and providing such hazardous spots with exhaust inlets and return points is very essential. It should however be noted that the provision of exhaust ventilation only is a limited measure and should be considered as supplementary to the general ventilation by dilution.

Natural ventilation, unless induced artificially, occurs through un-tight closures of doors, windows and various other openings by effect of temperature and

No.	Pollutants	Origin of pollutants	Effects
1.	Sulfur dioxide	Industries, especially where coal and oil are fired.	Irritation to eyes and respiratory system, increased mucus production, cough and shortness of breath.
2.	Carbon monoxide	Principal contributors are automobile exhausts and industries.	Reduction in oxygen carrying capacity of blood.
3.	Oxides of nitrogen	Automobile exhausts.	Pulmonary irritant, affecting lung functioning.
4.	Hydrocarbons	Automobile exhausts.	Lung cancer.
5.	Chlorine	Chloralkali industry.	Mucosal irritation.
6.	Ammonia	In the fertilizer industry and poultry farming.	Mucosal irritation.
7.	Hydrogen sulfide	Manufacture of coke, viscose rayon, distillation of tar and petroleum.	Illness and excessive inhalation leads to death.
8.	Acids and aldehydes	Chemical industries.	Eyes, nose, and throat irritation.
9.	Suspended particulate matter (SPM)	Industries; automobile exhausts.	Respiratory diseases.
10.	Dust	Industries and automobile exhaust.	Silicosis.
11.	Asbestos	Roofing, brake linings.	Asbestosis; is carcinogenic to workers.
12.	Lead	Automobile exhausts.	Cumulative poison; affects central nervous system.
13.	Beryllium	Aerospace industry; manufacture of household appliances.	Fatal lung disease; heart and lung toxicity.
14.	Manganese	Mining operations.	Damages nerves and reproductive system.
15.	Benzene	Automobile exhausts and manufacture of chemicals.	Leukemia; chromosomal damage.
16.	Pesticides	Manufacture of pesticides.	Leads to death if inhaled in excess; causes depression.

Table 9.
Major air pollutants and their effects on human health.

difference of densities of the indoor and outdoor air, velocity and direction of wind. **Figure 4** shows schematically the distribution of air pressure and the difference in the elevation of the inlet and outlet openings provided for natural ventilation.

The various types of natural ventilation are the airing and aeration. Airing is achieved by periodic opening of air wings in the windows. It is essential that airing will induce no changes in the temperature of air nor be conductive to the formation of mists, condensation of water vapors on the wall, floor or exposed glass surfaces. Aeration is the induced natural ventilation that serves the purpose of general dilution ventilation for interior space to maintain the necessary air quality within the specified limits. Aeration is effective only if the mechanical control mechanisms for frequent shutting and opening of the window wings are well designed and dependable. In the work areas where air changes cannot be effected too often, use is made of exhaust ducts or shafts the uppermost portion of which is led above the roof

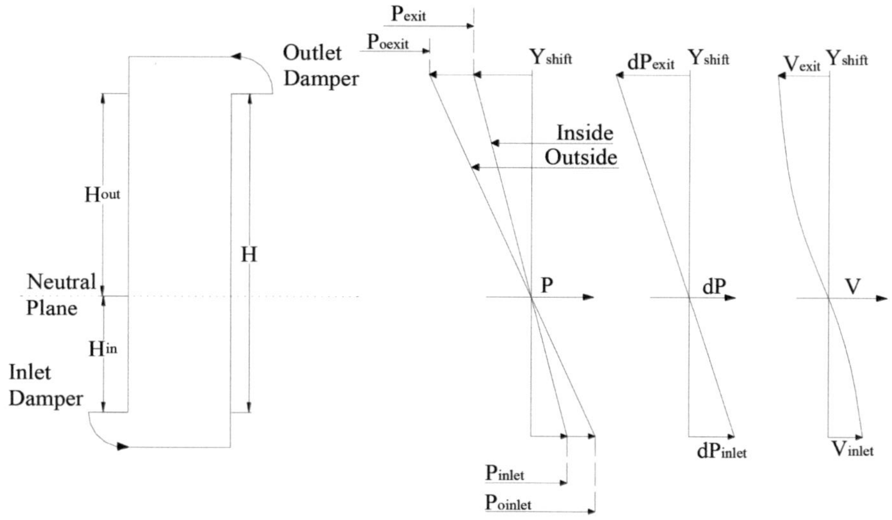

Figure 4.
Air motion pattern in natural ventilation.

ridge. To increase efficiency of air exchange the exhaust air ducts are crowned with diffuser attachments or deflectors.

Deflectors increase the draft of the indoor warm air due to wind force. There exist a large number of deflector types of various designs. The deflector is a branch pipe that flares at the top to form a diffuser which is capped with a baffle secured to the branch pipe upper portion. Both the diffuser and baffle are housed in a cylindrical shell which through arms is fixed to the diffusor so that certain space separates them. The deflector is installed over the top of an exhaust duct or shaft to catch wind. By wind action, air currents flowing about the shell outer surface build up a vacuum in the diffuser thereby increasing the exhaust flow from the enclosure.

Mechanical ventilation can be of different types, namely the forced (plenum), exhaust, and plenum exhaust ventilation. Forced ventilation is a system whose function is to force fresh air through ventilation ducts to work spaces with excess heat and minor concentrations of pollutants in the air. The removal of foul air through ventilation shafts is not only due to heat load and wind force but also due to the effect of overpressure created by forced ventilation. Fresh air is distributed between various zones of working spaces through a network of air ducts passing to various work places where it emerges from various inlet grills and other outlets. Exhaust ventilation can be used in enclosures free of releases into the air of harmful substances, also in work zones where few air changes are sufficient, in auxiliary type structures, service rooms and warehouses. In such cases, fresh air is supplied by natural ventilation through air vents and windows, through voids in the walls and floors, also through un-tight closures at doors, windows and from adjacent rooms. Plenum-exhaust ventilation is necessary in all work spaces where there are demands for an exceptionally dependable air exchange. When using this type of ventilation it is desirable to build up a little overpressure within work spaces where air contains negligible amounts of airborne contaminants to prevent the spread of vitiated air from the adjacent zones where the harmful contaminants may be quite significant. Fresh air outlets should be made so as to face the working zones. Hybrid or compound ventilation, the combination of mechanical and natural ventilations, can be resorted to where mechanical ventilation may unobjectionably be supplemented with natural ventilation for delivering or exhausting the air. Local ventilation may be of the exhaust and the plenum type. Local exhaust ventilation is intended for the removal of specific indoor air contaminants directly from where they form or release to prevent their spread throughout the environment of the working area, or to minimize harmful releases into the air in work zones. The advantages of local ventilation lies in that it prevents pollution of the entire air space within a working enclosure and exhausts minimum air volumes containing concerned amounts of dangerous impurities.

Air-conditioning serves to maintain a desired environment (air temperature, humidity, purity and velocity) within working zones regardless of the outdoor climatic conditions. Industries use air-conditioning in laboratories where work is measurable in high degree of accuracy and in departments whose output is high-precision instruments and devices. Air-conditioning provides for cleaning, heating or cooling, humidifying or drying of the air before admission to the work zone. Normally, air-conditioning systems use return air except for some production processes where recirculation of air is not allowed for health reasons. All processes of air-conditioning are controlled automatically. Self-contained, general purpose air conditioners are designed to control room temperature from +18 to +20°C, to within ±1°C. Other types of air-conditioners control humidity from 30 to 70%, to within ±5%. The choice for the use of return air is governed by reasons of saving energy. In summer the air conditioner requires too much cold water for air cooling, and in winter, hot water for air heating.

Ventilation equipment includes ventilation shafts, air ducts, ventilating fans, air cleaners and heaters. Intake (plenum) shafts provide supplies of atmospheric air. The concentrations of pollutants in fresh outdoor air supplies measured at air outlet points should not exceed 30% of their maximum safe concentrations. Air shafts can be separate structures or lean-to extensions. Plenum shaft intakes are normally grilled with louver-type shutters to protect them from atmospheric precipitation. The choice of a proper size for plenum openings and shaft sections relies on the air velocity which for openings ranges from 4 to 12 m/s, for ducts from 2 to 6 m/s. Air ducts located inside the building are intended to convey air to working zones or remove vitiated air from workrooms at points specified by the design of the ventilation system. The choice of construction materials for air ducts is made depending largely on the medium they are supposed to convey with due regard to the fire-prevention requirements. Air ducts can be round or rectangular in shape. Choice of the proper size or diameter of the air duct depends on the efficiency of the selected fan and rests on the assumption that the air velocity is high enough to prevent dust load. To minimize the air drag, the inner surfaces of air ducts are made smooth, and the branching, bends and transitions easy. Flange-type coupling ensures a tight connection and a joint that is easy and fast to assemble. The delivery of the air and its distribution at workspaces is affected by way of air outlets of various design, which are normally grilled branch pipes communicating with a common air duct.

Humidifiers are devices for maintaining desired humidity conditions of the air supplied to working zones. During the warm period they can also serve for cooling the air. Air heater is a device for conditioning the air to a desired temperature before admission to industrial buildings during the cold period. Fan is a device for delivering or exhausting large volumes of air or gas through a ventilation system with a low pressure increase. By manner of action, distinction is made between centrifugal and axial fans. The impeller consisting of a number of interconnected paddles or blades is an essential part of a fan of any type. Centrifugal fan has an impeller of a paddle-wheel form, a spiral jacket, a shaft, a pulley and bearings. The air enters axially at the center of impeller and is discharged radially by centrifugal force into the spiral jacket where it gets partially compressed. It is also called paddle-wheel or radial fan. Axial fan consists of an impeller or rotor carrying several blades of airscrew from working in a cylindrical casing, sometime provided with fixed blades. It is also called propeller fan. The air enters axially at the center of the impeller and is discharged centrally along its axis of rotation. The direction of the air discharge can thus be easily altered by changing the direction of the impeller rotation. Axial fans are usually driven by a direct coupled motor, but may be geared up or down through a belt transmission by choosing suitable pulleys. Axial fans are usually used for conveying large volumes of air against small counter pressures when friction or pressure loss in the ventilation system is not more than 0.1–0.25 kPa. Normally the choice for a fan depends on the value of total pressure loss which for low pressure systems is 1 kPa; medium pressure systems, more than 1–3 kPa; and high pressure systems, more than 3–12 kPa. Fans also bear numbers which, in decimeters, indicate the impeller diameter. The choice for a suitable fan can be made using fan performance curve usually given in ventilation equipment catalogs. Fan characteristics can be individual and universal. The individual characteristics show the efficiency (m^3/h) on the abscissae versus pressure (Pa) on the ordinate. Because the efficiency and the pressure produced by a fan depend on the speed of impeller (rpm) and friction loss of the air ducts the fan characteristic is constructed for various speeds and can be used against the various pressure loss values of the air duct. Usually, the same fan operating at a given speed discharges different air volumes, depending on the friction loss of the air duct network to which it is connected. When looking for a fan of a suitable efficiency, take the abscissae point of the necessary air discharge and find the necessary speed (rpm) and efficiency of the fan by constructing a vertical line till it intercepts the curve in a point across the ordinate. Power consumption (kW) of fans can be obtained from the expression:

$$N = \frac{VxP_f}{3600x102x\eta_f x\eta t} \tag{11}$$

where V is the volume flow rate, m^3/h; P_f is the air duct friction loss, Pa; η_f is the efficiency of fan; and η_t is the efficiency of transmission, which for direct-coupled motor is 1; and for V-belt transmission, 0.9–0.96.

5.1 Air cleaning equipment

One of the basic principles of ventilation of a building with pollutants is the cleaning of both plenum and exhaust air, if the last is heavily dust-laden. To clean the plenum air which normally has little dust, use is made of dry porous, wet porous and electrical filters. The exhaust dust laden air should be cleaned before it is exhausted to atmosphere. The process is affected by means of dust collectors or catchers in which dust separation is achieved by gravity force, centrifugal force, or by bag filters. If the cleaned air contains not more than 1–2, 40–50 , and more than 50 mg/m^3 of dust particles, cleaning is regarded as fine, medium-coarse, and coarse, respectively. There are many types of air cleaning equipment.

De-dusting chamber is an apparatus by means of which suspended dust can be precipitated; the process is generally affected by means of an exhaust fan. The dust-laden air through air ducts enters a chamber with a section several times that of the air duct in which the air loses its speed and dust separates from the air by pull of gravity. Only, coarse dust settles in dust collectors of this type which constitute the first stage cleaning for initially large content of airborne dust.

Cyclone is a type of a conical dust extractor in which dust is separated from the cyclonically rotating dust-laden air by centrifugal force. Airborne dust particles are forced against the walls of the dust separator, lose speed and settle down in the conical lower portion of the collector from which the collected dust is periodically removed. Cleaned air is exhausted to atmosphere through the upper pipe.

Bag filter is a widespread device for cleaning of the air. It is an apparatus for separating air from fine dust and constituting essentially of a canvas cloth through which dust-laden air is forced or driven by pressure of suction leaving dust particles on the outside of the canvas. It is equipped with a shaker device to regularly clean the canvas. To improve the removal of dust from the canvas filter, the shaker may be assisted by back-flushing, i.e., blowing clean air through the canvas from the inside. Baghouse filters produce a high degree cleaning (99%) when catching dust particles sized from 0.3 to 4 μm. Their disadvantages include large dimensions need for frequent removal of the collected dust and filter cleaning for, as the cloth resistance of the filter increases, its capacity lowers.

To clean the plenum air from fine dust, ventilation systems employ air cleaners with paper filters. The filtering medium is pure cellulose made in the form of very thin layers of crimp paper and packed into stacks of a definite thickness. Filter paper for quantitative purposes is treated with acids to remove all or most inorganic substances. The disadvantages of paper filters are relatively high resistance to airstream and frequent need for changing the filter paper stacks. Besides, paper filters are impracticable for high concentrations of dust as their resistance to the air passes increases, and this demands for immediate replacement as it reaches 120–150 Pa. The efficiency for fine dust separation is 92–95%.

Electrical precipitators are dust collectors in which the process of dust separation is affected by an electrical precipitation plant using a unidirectional electric field in which airborne dust particles are attracted to, and collected on, the positive electrode from which dust is shaken off into a dust bunker. These are also called electrostatic precipitators whose separation efficiency for fine dust is 95–99%.

The choice of a filter or dust arrester is based largely on the required degree of cleaning, content of dust in the dust-laden air, nature of dust and size of dust particles, efficiency and pressure of the ventilating unit, resistance and capacity of the dust separator. The physical properties of dust are also important. Thus, inertia-type de-dusters and baghouse filters are ineffective for sticky moist or fibrous kind of dust. Dust collectors, cyclones or electrical precipitators are usually ineffective for dusts in which the vast majority of particles are less than 10 μm in size. Content of dust suspended in the air is also important for choosing an air cleaner of a suitable type and capacity.

6. Conclusion

Pollutants of industrial indoor environment, along with importance of indoor air quality and health are elaborated in detail. The noise calculation charts are provided for interference of noise waves based on a benchmark solution. The grades and flag colors are notated to the noise calculation charts. Noise characterization calculations for indoor environment are presented using devised noise measurement equations. Indoor air quality and noise instrumentation based on source of noise are correlated and discussed. Effects of pollutants on human health are summarized. Human noise behavior along with psychophysiological measurements is briefly discussed. Ventilation systems along with air cleaning equipment for industrial indoor environment are discussed in detail.

Author details

Himanshu Dehra
Egis India Consulting Engineers Pvt Ltd., Gurugram, Haryana, India

*Address all correspondence to: anshu_dehra@hotmail.com

References

[1] Banerji SK. Environmental Chemistry. Second ed. New Delhi: Printice Hall of India; 1999

[2] Lin B, Huangfu Y, Lima N, Jobson B, Kirk M, O'Keeffe P, et al. Analyzing the relationship between human behavior and indoor air quality. Journal of Sensor and Actuator Networks. 2017;**6**:13. DOI: 10.3390/jsan6030013 Available from: www.mdpi.com/journal/jsan

[3] Vlek C, Steg L. Human behavior and environmental sustainability: Problems, driving forces, and research topics. Journal of Social Issues. 2007;**63**(1):1-19

[4] Stansfeld SA. Noise effects on health in the context of air pollution exposure. International Journal of Environmental Research and Public Health. 2015;**12**: 12735-12760. DOI: 10.3390/ ijerph121012735

[5] Dehra H. Acoustic filters for sensors and transducers. In: International Conference on Applied Energy (ICAE 2018); 22–25 August 2018; Hong Kong, China. Energy Procedia, Elsevier; pp. 8. Available from: http://www.applied-ene rgy.org/icae2018/

[6] Dehra H. A paradigm of noise interference in a wave. In: Internoise-2018, 47th International Congress and Exposition on Noise Control Engineering, Chicago, Illinois, USA; 26–29 August 2018; INCE; pp. 12. Available from: https://internoise2018.org/

[7] Dehra H. A unified theory for stresses and oscillations. In: Proceedings of CAA Conference Montréal. 2007, Canada, Canadian Acoustics. September 2007. Vol. 35. No. 3. pp. 132-133

[8] Dehra H. Chapter 5: Acoustic filters. In: Romano VA, Duval AS, editors. Ventilation: Types, Standards and Problems. New York: Nova Publishers; 2012. pp. 135-154

[9] Dehra H. Power transfer and inductance in a star connected 3-phase RC circuit amplifier. In: Proceedings of AIChE 2008 Spring Meeting; 6–10 April 2008; New Orleans, LA, USA; session 96a, pp. 7

[10] Dehra H. A novel theory of psychoacoustics on noise sources, noise measurements and noise filters. In: INCE Proceedings of Noise Con 16 Conference; 13–15 June 2016; Providence, Rhode Island, USA; pp. 933-942

[11] Dehra H. A slide rule for noise measurement. In: 10th International Conference on Sustainable Energy Technologies (SET 2011); 4–7 September 2011; Istanbul, Turkey; pp. 5

[12] Dehra H. On sources and measurement units of noise. In: Proceedings of International Conference on Innovation, Management and Industrial Engineering (IMIE 2016); 5–7 August 2016; Kurume, Fukuoka, Japan; ISSN: 2412-0170. pp. 219-227

[13] Dehra H. A paradigm for characterization and checking of a human noise behavior. In: WASET Proceedings of 19th International Conference on Psychological and Brain Sciences; 11–12 May 2017; Montréal, Canada; PISSN:2010-376X, EISSN: 2010-3778. pp. 658-666

[14] Dehra H. A guide for signal processing of sensors and transducers. In: Proceedings of AIChE 2009 Spring Meeting; 26–30 April 2009; Tampa, FL, USA; session 6b; 2009. pp. 9

[15] Dehra H. Acoustic signal processing and noise characterization theory via energy conversion in a PV solar wall device with ventilation through a room. Advances in Science, Technology and Engineering Systems Journal. 2018;**3**(4): 130-172, Special Issue on Multidisciplinary Sciences and Engineering. DOI: 10.25046/aj030414

[16] Dehra H. Solar energy conversion and noise characterization in photovoltaic devices with ventilation, invited chapter in book. Photovoltaic Materials and Solar Panels. United Kingdom: InTech Publication, IntechOpen. DOI: 10.5772/intechopen.79706. (Available from: https://www.intechopen.com/chapter/pdf-download/62929) (20 pages)

[17] Dehra H. Solar energy absorbers, invited chapter in book. In: Manyala R, editor. Solar Collectors and Panels, Theory and Application. InTech Publication; 2010. pp. 111-134. ISBN 9789533071428 DOI: 10.5772/10334. (24 pages) (Available from: https://www.intechopen.com/books/solar-collectors-and-panels–theory-and-applications/solar-energy-absorbers)

[18] Dehra H. Acoustic filters for sensors and transducers: Energy policy instrument for monitoring and evaluating holy places and their habitants. In: Taner T, editor. Energy Policy. InTech Publication IntechOpen; 2019. DOI: 10.5772/intechopen.81949. (Available from: https://www.intechopen.com/chapter/pdf-download/64590)

[19] Dehra H. A heat transmission model for a telephone line. In: 21st Canadian Congress of Applied Mechanics; 3–7 June 2007; Toronto, Ontario, Canada: Department of Mechanical & Industrial Engineering, Ryerson University; 2007. pp. 356-357

[20] Dehra H. A two dimensional thermal network model for a photovoltaic solar wall. Solar Energy. 2009;**83**(11):1933-1942 (10 pages). DOI: 10.1016/j.solener.2008.07.014. Available from: http://www.sciencedirect.com/science/article/pii/S0038092X08001862

[21] Dehra H. A combined solar photovoltaic distributed energy source appliance, natural resources. Scientific Research. 2011;**2**(2):75-86 (12 pages). DOI: 10.4236/nr.2011.22010. Available from: http://www.scirp.org/journal/PaperInformation.aspx?PaperID=5258

[22] Dehra H. A theory of acoustics in solar energy, natural resources. Solar Energy. 2013;**4**(1A):116-120 (5 pages). DOI: 10.4236/nr.2013.41A014. Available from: http://www.scirp.org/journal/PaperInformation.aspx?PaperID=29361

[23] Dehra H. A mathematical model of a solar air thermosyphon integrated with building envelope. International Journal of Thermal Sciences. April 2016;**102**: 210-227 (18 pages). DOI: 10.1016/j.ijthermalsci.2015.11.022. Available from: http://www.sciencedirect.com/science/article/pii/S1290072915003750

[24] Dehra H. An investigation on energy performance assessment of a photovoltaic solar wall under buoyancy-induced and fan-assisted ventilation system. Applied Energy. 2017;**191**:55-74 (20 pages). DOI: 10.1016/j.apenergy.2017.01.038. Available from: http://www.sciencedirect.com/science/article/pii/S0306261917300466

[25] Dehra H. Electrical and thermal characteristics of a photovoltaic solar wall with passive and active ventilation through a room. International Journal of Energy and Power Engineering. 2017; **11**(5):514-522 (9 pages). DOI: 10.5281/zenodo.1130291 Available from: http://waset.org/publications/10007024WASET (scholar.waset.org/1307-6892/10007024)

[26] Dehra H. The entropy matrix generated exergy model for a photovoltaic heat exchanger under critical operating conditions. International Journal of Exergy. 2008; **5**(2):132-149. Inderscience ISSN online: 1742-8300. DOI: 10.1504/IJEX.2008.016671

[27] Dehra H. A multi-parametric PV solar wall device. In: Proceedings of IEEE International Conference on Power, Control, Signals and Instrumentation Engineering (ICPCSI-2017); 21–22 September 2017; 978-1-5386-0814-2/17/$31.00 ©2017. Chennai, India: IEEE, pp. 392-401. Available from: https://ieeexplore.ieee.org/document/8392324/ (10 pages)

[28] Dehra H. Characterization of noise in power systems. In: IEEE International Conference on Power Energy, Environment & Intelligent Control (PEEIC2018); 13–14 April 2018. 978-1-5386-2341-1/18/$31.00 ©2018. IEEE, Greater Noida, India: IEEE; in Press. pp. 321-330 (10 pages). Available from: http://www.glbitm.org/PEEIC/index.html

[29] TSI Incorporated. Indoor Air Quality Handbook. A Practical Guide to Indoor Air Quality Investigations. Shoreview, MN 55126 USA. 2003

[30] Poltev MK. Occupational Health and Safety in Manufacturing Industries, Moscow: Mir Publishers; 1985